D1047184

House-Dreams

Other books by Hugh Howard

Wright for Wright

The Preservationist's Progress

How Old Is This House?

Bob Vila's Complete Guide to Remodeling Your Home

House-Dreams

The Story of an Amateur Builder
and Two Novice Apprentices and How
They Turned an Overgrown Blackberry Patch,
Ten Truckloads of Lumber, a Keg of Cut Nails,
and an Antique Staircase into a Real Home

HUGH HOWARD

ALGONQUIN BOOKS
OF CHAPEL HILL
2001

Published by
ALGONQUIN BOOKS OF CHAPEL HILL
Post Office Box 2225
Chapel Hill, North Carolina
27515-2225

a division of
WORKMAN PUBLISHING
708 Broadway
New York, New York 10003

Printed in the United States of America.
Published simultaneously in Canada
by Thomas Allen & Son Limited.
Designed by Amy Ruth Buchanan.

Library of Congress
Cataloging-in-Publication Data
Howard, Hugh, 1952–
House-dreams : the story of an amateur builder and two novice
apprentices and how they turned an overgrown blackberry patch,
ten truckloads of lumber, a keg of cut nails, and an antique staircase
into a real home / by Hugh Howard.
p. cm.
ISBN 1-56512-293-3
1. House construction—Case studies. 2. House construction—
Amateurs' manuals. 3. Howard, Hugh, 1952—Homes and haunts.
I. Title.
TH4815.H69 2001
690'.8372'092—dc21
00-054827

10 9 8 7 6 5 4 3 2 1
First Edition

For my wife,

BETSY,

and our daughters,

SARAH and ELIZABETH

CONTENTS

House-Dreams

PROLOGUE

THE MAYFLOWER

When the stagecoach topped the hill from Monterey,
and we looked down through pines and sea-fogs
on Carmel Bay, it was evident that we had come
without knowing it to our inevitable place.
—Robinson Jeffers

Major decisions often get made for small reasons. The day we resolved to build a house for ourselves demonstrates how the little can beget the big. The impetus for one of the most momentous decisions of our lives proved to be a tiny wildflower.

My wife, Betsy, and I were walking in the woods that spring day in 1992. Our one-and-a-half-year-old daughter, Sarah, rode happily in the child carrier on my back, kicking me in the ribs each time she exclaimed at spying something new.

We were walking there as prospective land buyers. For some years, Betsy and I had considered building a house for ourselves.

As we scanned the terrain that afternoon, looking for a suitable building site, my mind was busy with lots of questions. One of them was, *What will the house look like?* In the past, we had repeatedly discussed what we wanted, but no single resolution had been reached. Other questions that offered themselves were, *How many bedrooms? How long will it take to build a house? Can we afford to do it?* We knew I would be the builder in order to save money, but that, too, raised a question: *Do I, having never built a house from the ground up, have the skill and energy to do it and do it well?*

Since Sarah's birth, our lives had been largely consumed by the unfamiliar joys of parenthood. We knew that, despite our share of sleepless nights, it was a golden time, and we let the daily discoveries wash over us. Just days before our little walk in the woods, though, we had gained a powerful reason to rouse ourselves from our languor.

We had been talking about how cramped our house had begun to feel with Sarah's newfound mobility and her ever-growing array of toys. I was sorting papers on my desk while we talked, but something in Betsy's manner caused me to look up. Her face was impassive when she said, "With another child in the house, we will really feel the squeeze." She fell quiet a moment (thinking back on it now, I realize it was truly a pregnant pause). "We should probably think about another bedroom," she added.

Her pregnancy was news to me, though it wasn't unexpected. With an embrace and a few words, we reassured each other that we both felt having another baby was a fine thing.

The news had led me to follow up on overdue resolutions, to make the phone calls and have the conversations that led to our hike in the woods.

We walked beneath untamed evergreen trees, mostly hemlock and white pine, guided by the bright orange ribbons of surveyor's tape that marked the boundary. Preoccupied with worries about the cost per acre and property-tax rates, I only half saw what I was supposed to be examining. But Betsy, as usual, was paying attention to the details. She knelt down, peering intently at the ground.

She called to me, "What's this flower?"

I recognized it immediately as a mayflower, a plant my mother had pointed out to me on childhood walks in the New England woods. Also known as the trailing arbutus, it's in the heath family, which boasts some fancy members, including the azaleas and rhododendrons. Compared to them, the mayflower is the forgotten stepsister.

Betsy smelled the flower—the fragrance of the mayflower is captivating. We helped Sarah out of the backpack so she could smell it, too. Instinctively she reached out to grasp the delicate flower, but she was distracted with talk of its shape.

When Europeans settled in North America in the seventeenth century, the mayflower was commonplace. But no longer. Botanists believe its rarity is the result of the plant's dislike of disturbance. Lumbering and grazing—sins characteristic of the idealized pastoral past—rendered it scarce.

The sighting of that flower brought us to a decision about that piece of land in a way that monetary calculations could not. The mayflower's complex nature had a special appeal to me. The trailing arbutus was local, perfectly adapted and suited to its setting. It didn't come from someplace else, unlike so many other members of the heath family that were transported from their native soils in Asia. I liked the paradox that the woody-stalked creeper with the leathery, hairy leaves seemed such a natural and permanent part of the landscape, yet was so transient.

By the time we returned home, Betsy and I had decided independently to buy that land and to build our home there. In retrospect, I understand that we aspired to the condition of the mayflower. We wanted to create something local. I wanted to build a house that was adapted to the climate of the northeastern United States and to the microclimate of our hills. I wanted Sarah and her unborn sibling to feel as if we belonged to this place. Like the trailing arbutus, we wanted to set down roots and blend into our environs. And we wouldn't have to disturb the mayflower—it was well away from the heart of the property where we would build.

We sensed those wooded acres were our place, and only half knowing how much work it would be, we set about building our house and establishing our home there.

This book is about that process.

The Footprint

Life can only be understood backwards;
but it must be lived forwards.
—Søren Kierkegaard

The rumble of its diesel engine was audible before the truck appeared. More than a year had been required to take title to the mayflower property and to get ready to build. But on this July morning, walking along the rough cart road that was to be our driveway, I was going to meet the man who would dig the foundation hole for our new house.

Leo of Gardina Brothers Excavating was chocking the wheels of the flatbed trailer. He prepared to unload its cargo, a vivid yellow Caterpillar excavator, a machine Sarah liked to call a steam shovel. Leo didn't see me as he went about the business of starting the excavator's engine and backed the machine off the trailer. When the excavator was ready to go, he looked up to find me a few feet away.

Leo Gardina had been operating earthmoving machines for more than thirty years. He was solidly built, his hair closely cropped. Upon first meeting him, I read his features as having been formed into a perpetual harrumph. That summer morning, we shook hands in greeting and I offered some remark about the weather. Leo kept the pleasantries short.

"Where're we working?" he asked.

We walked in on the cart road. I got a copy of the site and foundation plans from the passenger seat of my minivan. After a quick look at them, Leo waved them off.

"They look good," he said absentmindedly. He was looking past me to the site where I had staked out the foundation. The wooden stakes were hammered firmly into the ground, each about ten feet beyond the corners of the foundation, just as a book had said it should be done. Green and yellow twine connected the stakes to form a grid that looked like a giant tic-tac-toe puzzle.

"Those won't do me any good," said Leo. "I need lime marks to go by. One cut with that thing"—he gestured in the direction of the excavator—"and those strings'll be a tangled mess."

My first instinct was to say, *But the book said . . . ,* only I thought better of it. The problem Leo described was obvious, and he wasn't interested in hearing about bad advice from a misguided book.

"I think I've got some plastering lime somewhere. Will that do?"

Leo shrugged.

I found the lime and jiggled a line of it around the perimeter of the foundation. Meanwhile Leo drove the excavator in on the cart road.

An excavator always moves slowly, lumbering into position on its treads. Mounted on one side of its tractorlike base is the cab for the operator. On the other is the great elongated arm that gives the machine the look of a giant praying mantis. That arm bends and telescopes, extending the bucket at its end and bringing it back again. The base of the machine swivels, giving the great arm a 360-degree range of operation. Leo silenced the engine and dismounted.

"Make sure the lines go beyond the stakes," he instructed. "I need the ones outside the hole the most. The others will disappear when I start digging."

Within minutes, Leo went to his machine and I took a seat on a tree stump. Leo started the excavator and maneuvered it into position beside one of the lime lines that extended beyond the corner of the foundation.

This was my first chance to observe an excavator at work. The machine is less mobile than a bulldozer, which relies upon pure power to drive its blade forward, moving whatever dirt, rock, and other debris lie in front of it. A bulldozer can open a foundation hole, but the *dozer*—no one in the excavating business ever refers to the machine as a *bulldozer*—is better suited to smoothing. It's driven back and forth, its horizontal blade cutting off high spots and filling in low ones, like a knife smoothing the frosting on a cake.

Unlike the dozer, which relies upon a primitive pushing motion, the excavator uses its reach to dig a hole. Although steam-powered machinery is virtually extinct, the antiquated name *steam shovel* is quite descriptive of what an excavator does: it lifts out buckets of debris from the hole in an underhand shoveling motion. Unlike the dozer, which requires a ramp at the side of the hole to push the fill up and out, the excavator works from above, positioned on the original grade outside the hole. The end result is different, too. While a dozer creates a trenchlike hole with ramps pitching into it, an excavator can efficiently cut a house foundation with steep banks on all sides.

Leo had positioned the machine so that its treads were aligned with the front of the foundation. By maneuvering the control levers, he extended the bucket and lowered it to the ground. Then he drew it toward him, like a giant hand, palm down, scratching rigid fingers along the earth. The arm gradually bent at the elbow, and the bucket collected the debris the fingers had loosened. Leo then lifted the arm, pivoted it away from the lime grid, and extended it to dump the contents of the bucket into a pile.

The hole soon assumed the shape of a trench, deepening a few inches with each pass. Beneath the thin covering of topsoil, a two-foot layer of clay was gradually revealed. Loose stones, embedded in the earth, clunked and clanged into the bucket. As the trench grew, so did the accumulation of soil, clay, and stones outside the limed lines. When the bottom of the trench exceeded two feet in depth, the material Leo was digging changed to a bed of gray, crumbly shale.

Shale resembles phyllo dough enlarged to a gargantuan scale. Like the phyllo dough that covers spinach pie or baklava, shale consists of many stratified layers. The passage of time and tremendous pressures had consolidated clay, mud, and silt into stone. When assaulted by the giant bucket, it came apart, layer by layer.

Leo dug through the shale, and the steady rumble of the diesel engine became little more than an undertone as the steel teeth of the bucket tore into the stone. When the teeth of the bucket gripped a fissure, the giant excavator arm would flex and the stone would give with a crack. And again: *scrape, groan, crack.* The dirt and clay in the pile began to disappear beneath a layer of shattered gray stones. With the removal of each bucket of shale, the trench grew deeper and larger.

The construction process was truly under way.

As Leo worked, a small pickup truck appeared on the rough driveway. Before I could see the face of the driver, the cautious driving style revealed his identity. It was Charlie Briggs, a neighbor, a friend, and the man who had been most responsible for Leo's and my being at work on that spot. He had sold me the acreage on Stonewall Road that, for more than a century, had been at one end of his family's large farm.

Charlie might best be summarized in this way: *Charles A. Briggs Jr., born 1912, Red Rock, New York.* Just his place of birth (which, not coincidentally, was also the hamlet where I was building my new home) and his eighty-plus years conveyed more about him than comparable facts could about most people. He emerged gingerly from his vehicle, his shoulders hunched. We greeted each other, shaking hands and remarking on the weather.

The farmhouse where Charlie grew up was a few hundred yards away. As a boy, he had walked to school along Stonewall Road. For almost a mile, the country byway had been lined on both sides with his father's pastures, fields, and woodlots.

The road drops down a steep hill to a small valley carved by a stream. On its banks, a few dozen souls, including some of Charlie's ancestors, had established the hamlet of Red Rock. Indian Creek had powered the cotton batting mill, the sawmill, the gristmill, and, in turn, the village's development at the beginning of the nineteenth century. By the Civil War, the town had three churches, and Red Rock's population reached the hundreds. But the railroad veered around the hills that shape the Red Rock

Valley, bypassing the town. Ironically, the broken promises of progress that soon saw the village half-deserted made it appealing to people like Betsy and me who, late in the twentieth century, sought to escape New York City, a two-hour drive to the south.

When Charlie was a youngster, his daily walk down Stonewall had taken him to Red Rock Road, which parallels Indian Creek, and to a one-room schoolhouse. It still stands, boarded up since its closing in 1945. Some twenty years before that, Charlie had completed the program offered there, graduating with an eighth-grade education.

As a young man, he had taken Stonewall Road to the same junction, but instead headed west. After a brief apprenticeship as a butcher and shopkeep at the A & P in a nearby town, he had opened a small shop of his own in Red Rock. The goods for sale included local produce such as eggs, milk, and bacon, as well as canned goods, flour, and other staples. There was also a smattering of hardware (nails and the like) and dry goods (fabrics and sewing supplies). He had run the store for almost thirty years before going to work at the post office.

The Fates seem to have conspired to keep Charlie a provincial man. In 1975, he had been scheduled to appear in his postmaster role on a local television show called *Total Information News*. For the broadcast he traveled to Albany, New York, about thirty miles away, but that evening a lightning strike caused a power outage. No one got to see him on TV.

Charlie was one of the first inhabitants we met when we moved into our starter house in Red Rock in the early 1980s. In real estate shorthand, we had bought a "handyman special," a ramshackle cottage on an absurdly steep and wooded hillside. The site was quiet and isolated—our desire had been for a weekend escape from the noise and density of Manhattan. But there was no level and sunny spot for a vegetable garden, so Charlie gave us permission to plant our squash and parsnips in a corner of an adjacent field. More than ten years later, we approached Charlie to talk about purchasing acreage for a new house.

Charlie and I watched Leo at work. His company was welcome, since I was little more than a spectator, and the regular movements of the excavator were becoming monotonous. The sight prompted Charlie to tell me about the building of his house in 1946 across Red Rock Road from his store.

He had built the home for himself and his wife, Helen, and they still lived there. "I cut down the trees to build the place myself," he told me. Then he paused. Charlie was an unhurried talker who always had time for little silences in his conversation, whether for emphasis or just for the sake of quiet.

"Did it with an ax," he resumed. "There were no chain saws then." I squinted a little, trying to see the thin, rugged, hardworking young man he had once been.

Charlie Briggs was one of those people whose word you took. He certainly had his foibles—over the years he had become a bit of a hypochondriac, especially as his body showed signs of wearing out. Too often he told the old stories again and again (including the one about building his own home). But no one was likely to accuse him of pretending to be something he wasn't. His weatherworn face, his small nose and eyes, his earnest expression, all were a reprimand to those who might try to fool him. He would look straight at you and then quickly away if he sensed he was in the presence of pretense.

On a dozen occasions, I had heard Charlie say, "I was born in Red Rock." He would offer the information solemnly, looking at the ground, like a much-honored veteran remembering his war years. He would follow that with, "Lived here all my life."

He would look up to see if it registered with his listener how long that had been. Though he was not a prideful man, Charlie's ties to his tiny hamlet mattered a great deal to him.

"That area was a pasture, you know," Charlie said, gesturing toward Leo and the excavator. When Charlie paused, I didn't fill the silence. More was coming.

"When I was a boy, the top of this ridge was the best blackberry patch around here. I would walk over with a pail and fill it up in half an hour. The blackberries were *this* big." He gestured, his thumb and forefinger gapped a walnut-sized blackberry apart.

After we had watched Leo for a few minutes, Charlie straightened a bit and moved to go. We walked back to his truck, which he had parked next to a couple of junk cars.

Charlie gestured at the cars and traveled back in time again. "Those were my nephews' cars." Pause. "They learned to drive up here."

Both of us were silent.

"I'll get them out of here for you," he concluded.

He extended his hand. I grasped it and he returned my grip gently. I looked back at Leo at work in what had been the blackberry patch as Charlie laboriously climbed into the cab of his pickup and drove off.

My presence on the work site was not required except for the occasional moments when progress was to be measured using a surveyor's tool called a transit. That device is essentially a telescope on a tripod. Leo had set it up while I marked the foundation perimeter with the lime. After Charlie left, I sat at the ready near the transit as Leo worked his machine.

The transit was safely out of the reach of the excavator's long arm, perhaps sixty or seventy feet from the foundation. The legs of the tripod were planted firmly, with the transit head mounted on the platform on the top. Built into the platform was a bubble vial like that found in any level. It's used to set the head correctly so that the transit telescope will pivot on a true horizontal plane.

The transit is indispensable for plane surveying, in which topographical features—the rising and falling of the terrain—are plotted on a plan as if located on a flat surface. A professional surveyor uses the same tool when marking off the metes and bounds of a piece of property for a real estate transaction. There's nothing new about surveying, as the ancient Egyptians mastered the basics some four thousand years ago and the same procedures have been followed ever since. It was once a prestigious profession; both George Washington and Thomas Jefferson were trained surveyors. I felt in good company learning to use the transit, though they knew it as a theodolite.

In experienced hands, a transit can do many things, among them measuring angles of inclination in degrees, minutes, and seconds, which makes determining areas and distances a matter of a few mathematical calculations. That's handy for the surveyor charged with mapping a piece of property, but that wasn't our purpose. All we needed to do was determine when the hole Leo was digging was the right depth.

To operate the transit, two men and a second device, the so-called sticks, are required. Little more than a small bundle of graduated rods that telescope to the desired length, sticks have calibrations marked along their length in feet and tenths of a foot. When one man positions the sticks, the transit operator peers through the eyepiece of the transit. Focusing on the sticks, he can read the height of the grade at the crosshairs.

Let's say that the transit operator siting the sticks at point *A* sees the nine-foot marker on looking through the telescope.

That means the base of the sticks is nine feet lower than the plane of the telescope. By taking another reading in a second location—let's say the sticks at point *B* read six feet—the surveyor knows that point *B* is three feet *higher* than point *A*.

In staking out the foundation, I had used an antiquated transit bought at an estate auction for fifty dollars. Unlike Leo's contractor-grade model—which probably cost more than ten times as much—my transit was rather inexact. The lenses on the telescope didn't focus perfectly, so there was some guesswork in getting within an inch or two. My sticks were homemade, consisting of two lengths of scrap wood with a twelve-foot tape measure fastened at intervals with bands of electrical tape. My transit was adequate for staking out, but when I looked through Leo's that morning, I realized for the first time how inaccurate mine was.

Before driving any stakes into the ground, I had established a point of reference, a sort of ground zero, which was the height that was to be reached by the finished foundation. This would be important to the appearance of the house on its site and to septic system connections. I had driven a nail into a nearby tree at the desired height. A knot of fluorescent surveyor's tape made it easy to spot from a distance. By resting the sticks on the nail and pivoting the transit to take a reading there—my recollection is that it said two feet exactly—I had a basis for judging the relative depth of any spot in the hole with respect to the height established as ground zero. This way we could determine how much more digging was required. If a depth of, say, nine feet was required and the sticks read two feet at ground zero, then the hole would have reached the desired depth when the sticks read eleven feet.

The process seemed simple enough. But then so is balancing a checkbook.

Despite the size and power of the excavator, digging a foundation hole takes time. Watching from the sidelines like a substitute waiting to get in the game, I was eager to give the transit a try. But there was little point in checking until the hole Leo was opening was within at least a couple of feet of the desired depth. His experienced eye told him he wasn't there yet, so he just kept working.

My attention had wandered from him when there was a loud *pop!* I turned to look. The rumble of the engine faded to a quiet hissing sound. I walked toward the hole and saw that a thick black hose on the forearm of the excavator was bleeding a brownish fluid.

Leo climbed off the machine and we watched the last of the hydraulic fluid dribble out. His jaw was set, his mouth a frown. He shook his head. "I need my wrenches from the truck," he said quickly, and walked away. A few minutes later he had wordlessly removed the ruptured hose.

"I'll go get a new one of these," he said, gesturing with the two-foot tube. For the first time since he had stopped the machine, he looked directly at me. He read the disappointment in my expression. The job was at a standstill barely an hour after work had begun.

"I'll be back as soon as I can," he offered, smiling a laughing, reassuring smile.

Then he was gone. The trench he left behind was the size of a small lap pool, nowhere near large enough for the foundation of a house, too shallow to accommodate the eight-foot-tall cellar walls. The earth nearby was stained with the hydraulic fluid that was gradually seeping into the clay.

My plan for that day had been to observe Leo and monitor the progress of the excavation with the transit. I did have other work to do: as the self-appointed designer of the house, I had more drawings to complete before my job as self-appointed builder could begin. But the unexpected quiet at the building site allowed time for what I like to think of as free-range maundering. Or, if you prefer, daydreaming. I'd found over the years that daydreaming my way through a project *before* I actually did it helped me anticipate obstacles, build confidence, and devise tactics for tackling the difficult bits.

The preceding months had been busy, so the unexpected downtime was especially welcome. I had worked hard to finish several pending writing projects in order to clear a block of free time to work on the house.

We also had a new baby, Elizabeth Anne, born in January. As a new baby always does, Elizabeth reoriented the life of the house to her elemental schedule of sleeping and feeding. More than a few times I had awakened in the night with a start and looked into her cradle. In the wan moonlight, her tiny face could just be distinguished in the tightly wrapped receiving blankets that Betsy had made for her. She was so exquisite, and yet, more than once, I found myself reaching out involuntarily, in a parental panic at seeing her so still and pale, feeling for her warmth and heartbeat. But she was proving to be a healthy, happy child.

Our compact house seemed like the perfect place to welcome a baby and to establish new patterns for the four of us. But the sense of intimacy that made those days after Elizabeth's birth so pacific would soon give way to a feeling of being too close and crowded. In an atmosphere of contentment and expectation, I had begun the execution of the architectural plans on my computer. Over the months, there had been meetings with Leo and a range of other tradesmen. Foundation contractors, masons, and heating contractors had prepared estimates, individual prices to be slotted into an overall estimate.

I would do the carpentry, plumbing, wiring, and cabinetry, and some other jobs, too. Determining those costs had required the preparation of materials lists, the result of lots of arithmetical computations. By figuring out areas and by examining and reexamining the plans, the number of sheets of plywood needed had been determined, along with the number of two-by-fours, two-by-sixes, two-by-eights, and other pieces of lumber required, as well as the number of squares of shingles and the linear feet of trim boards, moldings, and clapboards and the pounds of nails needed, along with the lengths of copper and plastic pipe, fittings, insulation, wire, plugs, switches, electrical boxes, gallons of paint, square feet of flooring, bags of plaster, sheets of drywall—and a numbingly long list of other things.

My responsibilities as a builder would include buying the right supplies, and in the right quantities. Then the piles of supplies would have to be cut, shaped, assembled, and fastened in place. That day, sitting atop a fresh pile of dirt, I contemplated the shift from designer and number cruncher to builder.

A daunting prospect? Well, yes and no. Betsy and I had done similar work before, though never on such a scale. In remodeling our cottage house a few hundred yards away, we practiced the learn-by-doing approach. The place had been a wreck when we purchased it a dozen years before, and over a period of years, we had transformed it during weekends and vacations into a house of a certain style and comfort. (Betsy joked it took me nine years to finish the house, while she could produce a baby in a mere nine months.) I had mastered some basic skills and gained sufficient confidence that I had moved on to build a couple of new, small freestanding buildings, one a single-room office, the other a woodworking shop. But this house was a much bigger project—many more square feet, with two and a half bathrooms, a full kitchen, and, we hoped, a high degree of finish and style.

When I celebrated my fortieth birthday the previous fall, the birthday blues hadn't struck me especially hard. But there were nagging worries about not being exactly young anymore. And here I was, taking on what was for me a giant building job, one that would require a great deal of hard, physical work over a sustained time. Twenty-four-foot fir two-by-twelves weigh a lot —in the range of a hundred pounds if they've been exposed to rainwater—and soon I would find myself hefting them into place over my head. I would have to hold them in position while I drove nails the length of a long finger, using a giant framing hammer. Doing that repeatedly would require a certain amount of strength, balance, skill, and stamina.

The physical work had yet to begin, but I was already wondering at the limitations of my body. On a few occasions recently I had awakened in the middle of the night with a numbness in the ring finger and pinkie of my writing hand, a discomfort that could be the symptoms of carpal tunnel syndrome. It's a common injury among carpenters. Periodic back problems had also visited me over the years. A simple regimen of daily stretching exercises had kept the discomfort under control while I was living the sedentary life of a writer, but I worried about the effects of long days of physical work, week after week. Would my aging musculoskeletal system betray me?

In the end, would all the effort produce a house that lived up to expectations? How long would it take? Had I managed the estimates well enough to be sure that we really could afford what we were aiming to build? What had been forgotten in scheduling, budgeting, and planning the house?

The building process—mechanical though it may appear—is

organic in the sense that it involves growth. One kind of growth is obvious: During construction, a house gets bigger as more of its elements are cut, shaped, positioned, and attached. It gets taller and wider as the tissues of the place, the walls and floors and ceilings, grow thick with such sinews as wires and pipes and insulation. A skin of exterior cladding and interior plastering is added.

Another kind of growth occurs, too, one that involves human understanding. Even veteran builders and designers experience surprises along the way. No two houses are the same; no matter how many detail and elevation drawings are done, the finished appearance of a place is still a bit unexpected. Every step in the process brings realizations. Ask any architect about his or her favorite surprise—in candid moments, all architects will recall arriving on building sites to see designs they had known intimately on paper, only to discover delightedly an unexpected interplay of light on an interior space or some fortuitous accident of site or materials that made the place better than anticipated. Don't ask them about the unpleasant surprises, but be assured that they happen, too.

I WALKED AROUND THE start of the foundation hole. Nearby sat the great wounded excavator, immobile, with its arm limp at its side. Orienting myself to the lime grid on the ground, I found the spot where the front wall of the living room would eventually stand, enabling me to take in the view as if looking out one of the windows that, in a few months, would be set in the center of that wall. It wasn't yet ten o'clock in the morning, but the July sun was already drying the subsurface moisture from the excavated soil. The view was due south—the direction the main façade of the house would face—and I had to shield my eyes from the sun's brilliance. To the west was our temporary electrical entrance, a pole with a meter pad and electrical box from which we would run long cords to power our saws.

A thousand lucky turns in the road got me to the top of that pile with that imaginary window framing my view. None was more important than the affection for books that my father passed on to me. My childhood home was filled with books, and the ritual of reading was ingrained early. When I discovered—much later, in my late twenties—how absorbing buildings could be, one of the ways I exercised that interest was by studying books about buildings.

The connection between books and houses for me has always been unbreakable. My student days included no courses in architecture, so my knowledge of architectural history came entirely from books. Several groaning bookshelves attested to hours spent in bookstores and especially secondhand bookshops. There were a couple of textbooks to compensate for my lack of formal academic training. There were heavy coffee-table books filled with glossy photographs. Many how-to volumes had proved invaluable in remodeling our previous home. Numerous pages had been stained by my dirty hands when I'd referred to the books partway through various fix-it jobs.

In a sense, the link between books and buildings begins with a monk named Gian Francesco Poggio Bracciolini. Working in the monastery library at St. Gall, Switzerland, he came across a manuscript. Actually, it was one among many, since Brother Poggio made his find in the early fifteenth century, a few years before Johannes Gutenberg perfected movable type.

The manuscript was a nearly complete copy of a treatise that had been written some fourteen centuries earlier. The name of the work was *De architectura,* although it has come to be familiarly known as *The Ten Books on Architecture*. Its author was Marcus Vitruvius Pollio, whose name is usually shortened to Vitruvius. A Roman architect, he wrote the book in Latin, though its text was sprinkled with a variety of Greek terms. Its probable date of composition is 27 B.C.

Vitruvius's masterwork is a modest little book. In the inexpensive reprint edition I own it is barely three hundred pages long. But after its rediscovery in the fifteenth century, *The Ten Books* quickly found an audience and became a paradigm for the Renaissance itself, as it was the only work on architecture to survive from antiquity. Poggio's contemporaries were just beginning to examine ancient buildings, and Vitruvius's book came to be regarded as *the* primer to use in making sense of the surviving ruins.

To this day, Vitruvius's work remains among the most important architectural texts of all time, exerting an incalculable influence on the monuments of the Renaissance. But the manuscript Poggio discovered was not accompanied by illustrations. No drawings have ever been found (although some later editors and illustrators have added plates that they regarded as Vitruvian). The absence of original artwork has only added to the book's popularity, since architects and designers ever since have been able to use the practical advice they found in *The Ten Books* more or less as they pleased. With no specifications, drawings, or references to buildings, the reader is largely free to make his own judgments as to matters of style and taste.

That was part of the appeal of Vitruvius for me, too. Standing on the pile of dirt that morning, I was reminded of his advice about bedrooms and libraries: they should have an eastern exposure, advised Vitruvius, "because their purposes require the morning light." A look to my left revealed a shallow hole where the library was to be; on the second floor, with the same exposure, would be the master bedroom. Vitruvius didn't stop there—his book has more advice for builders than "Dear Abby" has for the lovelorn. Dining rooms should have a southwestern exposure, he counseled, "because the setting sun . . . in all its splendor . . . lends a gentler warmth to that quarter in the evening." He recommended an overall southern orientation for structures in northern climes. One doesn't have to be a Vitruvian scholar to recognize the good sense of his words. That's why our dining room would face west, and the front façade south.

We might have oriented the rooms the same way even without having referred to *The Ten Books*. Many architects understand his dictums: A friend once told me of hiring Charles Gwathmey, the noted contemporary American architect, to build him a house. One of the first things the two of them did was spend a day watching the sun come and go on the building site, together with a pot of coffee, a bag of sandwiches, and a six-pack of beer. Every wise designer is a Vitruvian, consciously or unconsciously. His teachings have become received wisdom, even common sense.

With no coffee or sandwiches that morning, I went to find Betsy and tell her of the delay. She offered a few words of assurance ("No big deal, really, it's only a day"), and I retired to my computer to continue working on the drawings for our house.

LEO DIDN'T RETURN UNTIL the next morning, when he appeared carrying the replacement hose in one hand and the bright yellow box that contained his transit in the other. While he bolted the new part in position, I busied myself setting up the transit.

The repair completed, Leo replenished the supply of hydraulic fluid, then tested the excavator. He moved the bucket carefully at first but gradually put more and more pressure on the mechanism. After dragging it across the stone and removing a few bucketfuls of shale, he climbed out of the cab to check the hose for leaks. Finding none, he nodded an okay at me and resumed digging where he had stopped the day before.

Years of experience are required to master operating a piece of heavy equipment like an excavator. The basics can be learned more quickly, but efficient operation requires a practiced fluidity. In the case of an excavator, one reason is that the simple swing of the bucket is a pendulum motion, so compound movements of the arm must compensate for the curved arc, flattening the cutting motion to produce a flat-bottomed hole. That means a row of levers and a set of foot pedals must be operated with great skill. Despite the range of adjustments required to accomplish this, there are no jerky movements when an experienced operator is at the controls. Leo's concentrated expression was as constant as the flawless movements of his machine.

Periodically we checked progress. He would clamber out of his seat when he thought it was time to check the depth of the hole. With the excavator's diesel engine still clattering, we would communicate with hand signals. He would position the sticks at a deep spot in the hole and I would focus the transit on the sticks. The hairline would indicate the point on the sticks that was level with the transit, and the actual depth of the hole could be computed using our ground-zero reading. Then he would move to another spot that looked deep enough.

Leo was not concerned with fractions of an inch. With two or three readings in his mind, he would remount his machine and go back to work. As the morning waned, he reached the proper depth in some areas and moved to widen and lengthen the hole. I missed Charlie's company as I stood by, idle but available.

At one point I spent a few minutes examining Leo's transit. Leo's was more sophisticated than mine, the head twice as large and made of bright plastics. It had calibrations where mine did not, and a great sturdy fiberglass tripod. Like mine, it had a spirit level at the fastening point between the head and the tripod.

Hmmm, yeah, that's interesting, I said to myself, looking at the transit. *Leo's transit seems to have two bubble levels.* I looked more closely at the one I hadn't noticed before. The bubble was not precisely aligned between the hairlines.

I felt my facial features go suddenly limp and involuntarily I uttered some unrepeatable expletive. Standing an arm's length away, I rotated the transit head on its pivot. The telescope clearly dipped downward rather than remaining precisely level, which meant that the readings I had been giving Leo were wrong.

I waved to Leo to stop, shaking my head. He saw my grimace and shut down the diesel.

"What's up?" he called.

I like feeling stupid about as much as I like getting correspondence from the IRS.

"The transit's not level; those readings were all off," I told him.

He looked at me uncomprehendingly.

"I just discovered there are two spirit levels on the base. Mine has only one." The look on Leo's face telegraphed his understanding. He said nothing but walked with me to the transit. His gaze followed my finger to the uneven bubble. He swore.

He dispatched me to handle the sticks in the hole as he quickly leveled the transit. I obeyed his hand signals as he moved me from one spot to another so he could assess the damage done. We learned that one large section of the main hole was two feet deeper than necessary.

"So what do we do?" I asked.

The dirt and shale couldn't simply be pushed back into the deep area, Leo explained, because the loose fill wouldn't compress properly, leading to uneven settlement and eventually to cracks in the walls and floor of the foundation. Crushed stone would have to be brought in and packed to bring the base back up to the level we wanted.

"What does that mean in terms of time?"

"At least a day," Leo said. There would also be the added expense of the stone and some man and machine time to spread it evenly.

I felt like having a foot-stomping tantrum, but Leo gave me no chance. He just shrugged and went back to work. By the end of the afternoon, he had finished digging the rest of the hole to the proper depth. And I had become a little smarter about the transit.

A FEW DAYS LATER, the foundation contractor arrived with his crew before 6 A.M. A load of crushed stone soon filled my gaping miscue, bringing the hole back up to its proper level. By midmorning, the forms were in place. They were low wooden structures nailed and braced to box in the concrete. The end product would be a two-foot-wide and one-foot-deep solid mass at the perimeter of the foundation, with steel reinforcement bars (called rebar) through the core. This footing would anchor the foundation walls. The concrete truck arrived and pumped a mix of water, cement, sand, and gravel down a chute into the forms at the bottom of the hole. The crew leveled its top surface, and by noontime, both truck and crew were gone.

Two days later, the men returned to strip the footings and assemble the forms for the walls. That took almost a full day.

Again, cement trucks came and went. The cement set hard in several more days, and the foundation crew came one last time to strip off the forms. Then I paid the bill, writing a check for slightly more than fourteen thousand dollars.

Handing over the check, I felt a little twinge of anxiety in my stomach. The sum came as no surprise, since it was consistent with the amount cited on the estimate prepared weeks earlier. But spending fourteen large, as the bookmakers of the world like to say, for a few tons of concrete that would shortly be buried left me mumbling to myself.

Betsy arrived at the site to check progress, Elizabeth on her hip. We walked around.

"Writing that check made me a little nervous," I admitted.

She looked at me quickly. "There's money to cover it in the account, isn't there?"

"Oh, sure, we moved some from that money fund of yours last week." About half of our building budget was based upon money Betsy had inherited from her parents, both of whom had died within the preceding three years. She remembered the transaction and nodded.

"No, I'm thinking about the rest of the process. You know, the hundred or two hundred checks we'll have to write to pay for the house."

"Will we have enough?" she asked.

If I had relied only on the public record to answer, the reply would have been a simple yes. When we applied for a building permit, an estimate of building costs had accompanied the application. The numbers were based on estimates and included the actual price quoted by the foundation contractor, $14,300. The Gardina Brothers had given me an hourly rate ($125) and an estimated number of hours (three eight-hour days), with the understanding that the time required could go up or down. My materials lists had produced estimates for framing materials (about $14,500) and the roofing ($2,500), as well as plumbing, electrical, and other jobs.

Thus, the estimated total cost of the house was a very modest $85,000 or about $35 per square foot.

What Betsy might have asked was, *How real are our numbers?*

Based on her inheritance, our savings, and our anticipated income over the period of construction, I felt we could cover the eighty-five large. Since my estimate was based on the best information available, I was able to hand the paperwork to the building inspector with no feeling of guilt or deceit. He glanced at the total, used it to calculate the fee ($262), and issued our building permit.

Yet all of us knew that those numbers were optimistic. Building projects do come in on budget—about as often as lottery tickets pay off big. We had never done this before, so our estimated cost reflected an ignorance of many questions we should have asked. We didn't know all the pitfalls, not to mention how to avoid them.

I walked Betsy to her car. I shrugged in answer to her will-we-have-enough question. "It'll cost more than we think, of course," I told her. "But I think we can manage it."

After Betsy drove off, I took a last walk around before starting for home, passing the three abandoned cars that Charlie had promised to have removed. One of them was a pre–World War II station wagon. A week earlier, the contractor who was estimating the cost of drilling a well had commented on it. The moment he had set eyes on the rusting hulk, he had stopped and stared intently at the wreck. The window glass was gone except for a few shards on the ground nearby. The seat upholstery had rotted away and the engine had been removed. Not a flake of paint remained, but the contractor looked at that sad old vehicle as if it were new. He peered through the gaping passenger-side door and shook his head.

"That was my first car," he said softly, still staring at what was left of the interior. "It's a 'thirty-four Chevrolet, just like mine was."

He walked around the car, examining it carefully. His face was impassive but I'm sure his memory roiled. Perhaps his thoughts that day were of the back-road exploits of a teenager testing the limits of his automobile; maybe he was enjoying backseat recollections of early sexual explorations with a girlfriend.

As I resumed my walk, I noticed a small pool of water in a deep dent on the car's hood. It had rained in the night. Then another movement caught my eye. Well beyond the car was a small wetland at the lowest point of the property. We had thought vaguely of putting a pond on that site someday.

There, beside the oversize mud puddle, stood a bird, tall and of uncertain coloring. It looked gray, though with a slight bluish cast. Its legs were spindly; its yellowish beak was long. Its neck was the most striking from my vantage, shaped as it was like a great S. The bird stood silently for a few moments. I remained motionless, but it turned and tilted its head toward me, as if to see me from a different angle. I realized it was a great blue heron. I had seen herons in photographs, but never in the wild.

With a strange leap, the bird took flight. Its neck remained crooked into an S and its ellipsoidal body looked heavy. It flapped its wings vigorously, seeming to hover at first. I wondered for a moment whether the bird might need to drop its thin legs back to earth to push off for added lift, but as it gained speed and altitude, the bird's flight was more fluid. It wasn't quite graceful, but there was an ease about its progress. Its cruising speed was surprisingly rapid as it turned in a great arc and flew overhead. The bird was no longer so ungainly, its wings waving purposefully. All that great blue heron had needed was to get airborne.

I glanced back once more at the topless concrete box in the large hole. The time had come when I was to go from bystander and bill payer to worker. There were worries, but I suddenly felt confident that the progress of construction would take care of them. Like that heron, all I needed was to gather some momentum.

Building the Box

If you want work well done, select a busy man:
the other kind has no time.
—Elbert Hubbard

A tall, slim young man arrived from Scotland at the end of July to become my helper. He appeared with a knapsack, a duffle bag, and a limitless curiosity about things American.

Betsy had been an occasional carpenter's helper as we had renovated our cottage, but that was before we had children. The two of us had agreed that on this project, Betsy's role in the construction would not be at the carpentry stage but would come later. She was an expert tiler and painter and would be in charge of those tasks. She would also be the head buyer, seeking the right appliances, fixtures, and other goods and materials at the right price. The two of us would make all major decisions as a team; that had worked well in our cottage. But the experience of the cottage had taught me that building is a job for two full-time workers. Enter Mark Lynas.

I had asked around about local boys. Several already had construction jobs. A couple of others didn't bother to return my phone calls. I didn't pursue them—who needs unreliable help? Before I had a chance to widen my inquiries, a friend telephoned. We were chatting about other matters when, quite by chance, she mentioned that her great-nephew was arriving to spend the summer and was looking for work to do. Or was she looking to keep him busy? I forget.

"Has he ever done construction work?" I asked. From the other end of the phone came the telephonic equivalent of a shrug ("Errrrrr . . ."), followed by the information that Mark was a sturdy lad, in his second year at the University of Edinburgh in Scotland. "I'm sure he's a capable worker," I was told. In little more than a moment, an understanding was reached. This unseen lad was suddenly coming to work for me and to stay with our family for a big chunk of the summer. Somehow, it had seemed the right thing.

At the train station, I wasn't so sure. Mark didn't look very sturdy to my eye. But then I'm pretty tall and lank myself, so we made a good pair. More worrisome to me was the discovery that his construction experience was close to nonexistent. Although he had helped his father lay up some masonry block a few years before, working with wood was utterly foreign to him.

His studies weren't much help either. His emphasis was political history, and he wanted to be a journalist. As a sometime journalist myself, I remarked offhandedly that he might want to try a real job first.

He kept his rejoinder to himself. At least he was savvy.

My plan called for Mark and me, over a period of about six weeks, to frame the house atop the newly completed foundation. The lumber and supplies needed to get under way had been ordered, and just before eight o'clock on the morning we were to begin, a red truck arrived, weighed down with lumber. Mark and I were at the house site to greet it.

Most American houses are wood-frame because the material has always been plentiful here. Unlike in Europe or Japan, where wood is both more rare and more expensive, there's lots of it to be had in the United States.

Construction-grade lumber is softwood, typically fir or pine. Our load that day consisted of a large stack of western fir with a few lengths of pressure-treated pine two-by-six. The pressure-treated stock, which is permeated under high pressure with an arsenic-based preservative to make it decay-resistant, was to become the sills. Sills are the wooden members that sit immediately atop the foundation. That's the part of the house that is most likely to be damp, which is why pressure-treated lumber is used.

The driver asked me where the load was going, and I indicated a space close to the foundation. The backfilling had been done, bringing the grade to within two feet of the top of the concrete wall.

"How about right there, next to this end of the foundation?" I suggested. "That's where we're starting."

"Fine with me. You want me to dump it?"

Dump it? I had been anticipating unloading the wood by hand. That's the way I had always done it before. But those deliveries had been less than full loads, while here our two-hundred-odd lengths of lumber filled the truck bed. And the offer was to off-load it in a single mechanical motion.

"I guess there's not much risk of breakage." It was a question, but I said it as if I knew what I was talking about.

"Nah, none of that two-by stock should break."

The decision was made, and he backed his truck into position and engaged the hydraulic mechanism that tipped the truck's horizontal bed. The end of the truck bed near the cab elevated slowly. When the flatbed was pitched about twenty-five degrees above horizontal, the lumber started to slide, the cue for the driver to release the clutch and drive forward. One end of the load hit the ground gently. He timed it deftly so that a moment later the other end of the load dropped off the back of the truck with a very loud *clunk*. The driver got out and handed me the order form to initial. "If any of it snapped, set it aside," he called over his shoulder as he got back into his truck. "I'll give you a credit next time I make a delivery." Then he was gone.

Mark and I looked at the stack, which was held together with strapping—metal bands that had been fastened like a belt around the wood. "Grab the snips from the toolbox in the van, Mark."

He looked confused.

"They're like oversize stubby scissors with orange handles," I told him. In a moment he was back from our family minivan with the right tool. I never had to explain to him again which were the snips.

We hadn't had to unload the wood by hand, but I was quickly reminded that for every tailwind that aids your progress, there's a shift that slaps you in the face. As we freed the wood from the strapping, I noticed for the first time that the pressure-treated two-by-sixes were on the bottom of the large pile. Since we needed those first, we would have to move the wood anyway. "So much for lucky breaks," I said.

We could have pushed the pile over, but on a work site, organization is everything. A tipped-over pile of wood poses a safety hazard—if you walk on it, you risk spraining an ankle when your weight causes it to shift and you stumble. Monitoring inventory is essential to anticipating what you need to order to keep busy, meaning time would be wasted counting what's left in a random pile. So we moved the pile about four feet to one side.

It was a laborious job that required ten minutes of repetitive work—lift, balance, walk two steps sideways, bend, set down the piece, square the pile, then do it all again. We were also learning each other's styles. Mark observed what I did and patterned his movements to mine.

Next I equipped him with a tool belt. A hammer hung through a loop on one side; a utility knife slid into a built-in sleeve on the other. A tape measure fit neatly into a pouch at center front. Two pockets on each side would hold a miscellany of nails and other small tools, depending upon the job to be undertaken on a given day.

Our next job was to attach the sills to the top of the foundation. The two-by-sixes needed to be cut to length, have holes drilled in them for the long bolts that pointed skyward from the foundation, and then be fastened firmly in place with nuts and flat washers. The sills had to be positioned precisely so that each corner was ninety degrees and square in relation to the others.

"Let's take some measurements, Mark." He took the body of my twenty-five-foot tape measure. I held the end of the tape over one corner of the foundation and gestured for him to go to the opposite end. The strip of yellow metal sat neatly between us along the top of the concrete wall.

"How long is it?" I asked.

He studied the tape for a moment before answering. "Fifteen . . . fifteen something."

"Fifteen foot what?"

"About fifteen feet plus this much." He held up his hand with the thumb and pointer finger about two inches apart. "Looks like two more small units and a couple of lines." After another pause, he concluded with a nod. "Fifteen feet two inches. And a bit."

I must have looked confused because he laughed at my expression. It was a slightly embarrassed laugh.

Leaving my end of the tape still clipped over the end of the foundation, I went down to look. "Looks like fifteen foot two and an eighth inches."

He nodded. "Okay," he said. "I'll know next time."

It dawned on me that Mark had never used a tape measure calibrated with feet and inches: his mind measured in metric. I wondered for a fleeting moment whether I shouldn't have looked harder to find a journeyman carpenter rather than a journalist collegian.

My formal training in the construction trades consisted in its entirety of two summers during college spent working in a furniture factory as an electrician's assistant. I bent what seemed like miles of aluminum electrical pipe, wired hundreds of fluorescent light fixtures, and helped my boss troubleshoot when electrical problems shut down machines. All of it was new to me.

I learned a few lessons about electricity. One was, *Be careful, but don't be afraid.* Another was, *The right tool used properly makes a job easier.* A third was, *All problems can be solved.*

My first summer as an electrician's assistant, I discovered that installing wiring wasn't as difficult as my calculus course the previous semester. Mastering the basics of electrical work required patience, hard work, common sense, and the willingness to ask endless questions. Sometimes, too, tasks had to be done several times to get them right, but all of it gave me a can-do feel for construction work. Hey, if I could learn wiring, why not carpentry, plumbing, and masonry?

Another discovery was how much I enjoyed my co-workers at the factory. They were blue-collar guys, very different from the profs, preps, and hippies at the private college I attended outside of Boston. My hair was long (this *was* early in the seventies) and many cars in the factory parking lot bore bumpers stickers that had messages like AMERICA, LOVE IT OR LEAVE IT. But my co-workers treated me as one of them as long as I did my job and checked my college-boy airs at the time clock when punching in each morning. Then I entered—and was welcomed into—an environment previously unknown to me.

I look back on it now as Chappy's world.

Chappy was a veteran of the conflict he always called "WW Eye-Eye." When he talked about his days building bridges and accomplishing other logistical feats in "Nort' Africa," it was as if he could hear the roar of a cheering crowd. The look on his face was distant before his features broke into the inevitable laughing smile.

Working with him, I couldn't help being impressed by his cleverness. He was resourceful—the absence of a certain tool or a needed replacement part was always a good excuse for him to improvise. He had a genius for keeping outdated machinery running even when it was long overdue for a one-way trip to the scrap heap.

What you couldn't tell from working with him was that until he had gone to war, he had been classified as irremediably slow. His teachers and parents had told him over and over again that he wasn't going to amount to much because he couldn't learn to read and write. "Must be retarded," they had said. "Poor kid." But Chappy had no shortfall of intelligence.

One day at lunch a group of men were talking. I was eating

my sandwich nearby, Joe College who listened a lot but didn't contribute much. Somebody asked Chappy, "When you built your house, Chap, you did the plumbing, right?" That question led to another and pretty soon Chappy was giving advice about laying out drain lines and vent stacks.

I was still back at the original question, absorbing its real implication. The man had said, "When you built your house . . ."

A couple of discreet inquiries among my co-workers confirmed that Chappy had indeed built his house. He was the carpenter, plumber, and electrician. He did everything.

That struck me as amazing. My father was worldly and successful—he was a lawyer with a busy practice, had held elective office, and was a man with a certain standing in our community. But he was utterly devoid of mechanical skills. Perhaps he was capable of performing hands-on tasks, but no one could remember seeing him so much as change a lightbulb. That had been left to my mother, who fixed things and had tool sense. For my part, I had set a few goals for myself, but *building a house?* That was a goal I hadn't even thought of.

That summer at the factory I learned that the rudiments of electric wiring could be mastered in a few short weeks of apprentice work and that doing physical labor agreed with me. Most important, Chappy demonstrated that one guy could build a house for his wife and daughters. Even a guy whose teachers said he was a dunce.

ATOP THE SILLS, WHICH lay flat and square and had been bolted to the foundation, Mark and I set the floor joists. At sixteen-inch intervals, the joists were to stand on their edges, boxed in at their ends by identical members that, by virtue of their position, are called headers. Like a giant rectangular rib cage, this frame would form the wooden platform from which the house would rise.

Building the platform was more difficult than it sounds. The joists had to reach across the cellar, spanning more than twenty feet. Because the ceiling of the basement was to be eight feet tall, the joists had to be positioned working from a stepladder.

We would crown them first, one of us looking along the length of the piece from one end to the other. Usually one edge rose slightly toward the middle. This slight arch (or crown) was to go upward, and we'd mark it with a penciled arrow. Over time, it would flatten from the weight of the floor and its load.

Kneeling on the foundation wall at grade level, I would balance a joist while Mark walked up a stepladder positioned on the foundation floor at the opposite wall. The long, awkward, and heavy length of fir rested, crown up, on his shoulder, braced with one hand while he climbed the ladder, balancing himself with his other hand. When his shoulders were even with the top of the foundation wall, he would slide the heavy joist onto the sill and muscle it into its precise position, as indicated by pencil lines drawn on the sill. Once we had nailed our respective ends into position, we would repeat the process with the next joist. Periodically we would switch jobs.

Not all the joists were the same length. Since the chimney and staircase were to rise through the first-floor platform, openings had to be framed for them. That meant some of the joists had to be cut short. We used specially designed U-shaped joist hangers of galvanized steel to suspend the joists from double-thickness headers, which acted as beams and were made of the same two-by-twelve stock of which the joists were made. They, in turn, were nailed to full-length joists with beam hangers.

All this required some acrobatic joist walking. Since the platform consisted only of 1½-inch-wide joists with 14½-inch gaps in between, a misstep would mean a nasty fall. Next, blocking had to be added—scrap pieces cut and nailed between the joists at roughly ten-foot intervals—in order to stiffen and strengthen the platform.

Once the joists, headers, rough openings, and blocking were in place, the plywood subfloor was then applied. This ¾-inch-thick layer capped off the foundation. From a distance, the flat wooden platform looked like a raft at a swimming hole, seemingly afloat over the work site.

Building the platform had required several days of demanding physical work. The labor left almost every muscle of my body sore and tired. I'm no masochist, but that tired-to-the-bones sensation was gratifying, especially since, at the end of each day, we inspected what we had done and could see genuine progress.

Back at home I found, for the first time in several months, that I slept soundly through the night. Middle-of-the-night awakenings had become familiar, periods during which I would mentally redesign the building, or worry and wonder about the process. Indeed, after a hard day's physical work, a deep, undisturbed sleep left me rested in the morning and ready to go to work. As one day of framing was succeeded by another, the aches lessened and my muscles loosened more quickly. Mark and I worked hard together and I liked the cycle, the sense of accomplishment, and the partnership, too.

No doubt the enjoyment would have been much less if I had been apprenticing for a long career as a builder. But my job as a writer already gave me pleasure, a reasonable degree of independence, and an adequate income. Building was just an avocation, but a week into the process, it suited me, just as it had other amateurs before me.

The Californian poet Robinson Jeffers once wrote, "If

anyone was ever bored, which is incredible, let him get five acres and grow a wood on them, and produce a stone house and twins and a book of verses." I'm no poet nor the father of twins. I was working with wood rather than stone. But I knew exactly what he meant.

In the year between our discovery of the mayflower and breaking ground for the house, I had visited northern California for the first time. After completing some business in San Francisco, Betsy and I had a few days to explore the coast south of the city, together with Sarah, who had just taken to toddling. At the recommendation of a friend, we spent one night in Carmel-by-the-Sea. A short entry in a guidebook enticed us to Tor House, an eccentric set of stone structures that had been home to Robinson Jeffers and his family.

In the preceding ten years, I had visited hundreds of old houses and written about many of them. But I was not prepared for Tor House and its sister structure, Hawk Tower. The home was based on a Tudor barn, and the tower on medieval Irish prototypes. But the impact of the buildings was more gut-level than intellectual. There was something immediate and even primeval about entering the stone portals and walking into the low-ceilinged rooms within. The builder was a man of words, but the language of Tor House was stone.

In a sense, the three of us took the tour that day. More precisely, Betsy and I took turns: one of us would view some of the interior with the Tor House guide while the other remained with Sarah, who was fascinated by the flowers and the butterflies outside. We learned how Jeffers had built the house for his wife and children.

To anyone with a dream of building his or her own home, the

tale of Tor House is a winning story. But Robinson Jeffers (1887–1962) seems to have been both a genius and a bit of a misfit. By age seventeen, he had earned a college degree; in the next handful of years, he pursued studies in literature and then medicine but found neither to his liking. Next he tried forestry, but that didn't take either: he found the new discipline too concerned with the lumber business and too little interested in explaining why there are redwoods, as he later wrote, that "neither grow nor grow old." Only when he married, moved to the outskirts of Carmel between Big Sur and the Monterey Peninsula, and purchased property on a promontory just south of town with a broad view of the Pacific did he find what he had been looking for.

He apprenticed himself in 1919 to a local builder named M. J. Murphy. Together they built the original stone house, a small cottage with three rooms on the ground floor and a loft space above. Jeffers and his wife, Una, named it Tor House after the Saxon word meaning high, rocky hill. When they began, the site was indeed a tor, a tall, stony bluff with no structures. It was in that house they brought up their twin sons.

Between 1920 and 1924, Jeffers himself built Hawk Tower, a four-story Irish tower. Having exhausted the supply of stone on the immediate site, Jeffers rolled stones from the shore up the slope to the building, self-consciously employing inclined planes as the Egyptians had done four millennia before. "Sea-orphaned stones," he called the native boulders that were his building material, some of which weighed as much as four hundred pounds. Later, he built a stone garage and an addition to the house. A low stone wall around the perimeter of the property was next. He landscaped the site, planting countless trees, framing the unobstructed vista of the Pacific Ocean, less than fifty yards away.

Poetry *and* stones proved to be Jeffers's chosen media. Tor

House is evidence of both. "[M]y fingers had the art / to make stone love stone," he wrote years later in his poem "Tor House." About building the house, Una wrote, "There came to him a kind of awakening such as adolescents and religious converts are said to experience." For Jeffers, the stone house was as much a part of his life's work as was his writing, and the two were inseparable.

The visit to Tor House inspired me to read a good deal of his writings. I admired most the poems and letters that concerned his life on that rocky point. It all seemed so very real and tangible—such as the detail that Hawk Tower got its name from the birds that liked to perch there during construction. Jeffers was a rugged outdoorsman, a private man who said of himself in late middle age, "I still try to avoid meeting people." He found pleasure and even good company in the stones he argued into place.

That romantic site, the visceral impact of the house, the fact that both he and Una died there and that the property is much as it was in his day, and even the eccentricity of the man and house, all contribute to the uniqueness of the place. But it's a house that appeals in particular to anyone with a house-building fantasy. Over a fireplace Jeffers built in Hawk Tower is a motto from Virgil. It's in Latin and reads *Ipsi sibi somnia fingunt*. In English, the phrase translates, "They make their own dreams for themselves."

Robinson Jeffers, my old co-worker Chappy, and our neighbor Charlie were certainly odd bedfellows. Different sorts of people build their own houses: theirs is a disparate fraternity. Jeffers was a master wordsmith; Chappy, profoundly learning disabled. Charlie was a proud but provincial man. But we all shared a common aspiration.

By the time we completed the platform, I realized that finding Mark had been a wonderful bit of serendipity. If he had been a tree, he would have been a birch: pale, destined forever to be thin, but much tougher than he seemed at first. He liked the work of framing. There's no downtime in framing, where you wait about for someone else to finish something. That appealed to Mark, since he had a strong dislike of idleness; he was bored by too little to do. That's a finer quality in a workman, I've found, than experience. He would prove an excellent co-worker.

We were finding Mark an agreeable guest in our home, too. He had the famous British reserve—on first meeting, I thought I read in his serious demeanor and few words that he had a disapproving air. As he grew more at ease with us, his funnier, warmer character came through clearly. A few days after his arrival, he discovered Sarah's bookshelves and admitted to me in a surprised way that he had yet to find the same joy in adult books that he had in children's literature. Sarah immediately recognized the spirit of the child in him, and they often read together. He delighted in quoting from Beatrix Potter: on putting his dirty work clothes into the washing machine one day, he remarked to Betsy, "Mrs. Tiggywinkle would be affronted!"

As comfortable as we soon became with having Mark in the house, there were reminders that we came from different places. For example, as a child of America, I found throwing things—and catching them—was quite natural, a skill all of my friends mastered early, thanks to games like basketball and baseball. I took it for granted that when you were ten feet up a ladder, you could call down, "Hey, Joe, toss me up that wrench, will you?" and expect that a moment later the tool would be lofted gently upward. You'd reach out, it would drop into your hand, and you would put it to use, having saved the trouble of climbing down.

One day while framing the deck, it was Mark who was up a ladder. Watching from below, I realized he needed a particular tool.

"Here, Mark, try this," I said, tossing a small nail puller up to him.

I caught him off guard. He flinched, almost falling off the ladder. Like a bear beset by bees, he batted the airborne tool away.

He looked down at me, speechless, the stricken expression on his face seeming to say, *Why are you throwing tools at me?*

I couldn't help laughing.

He was flustered, but he quickly defended himself.

"You *Americans,* I know, are preoccupied with throwing a ball around. In England, we *kick* a football around. That's more the proper thing."

He didn't like being laughed at. "By *football* I mean a *soccer* ball," he added sharply before I could wipe the grin from my face.

I stooped to pick up the tool he had unceremoniously knocked away, and climbed up the ladder behind him in order to hand it to him directly.

From then on, one of Mark's nicknames was Old Soccer Hands.

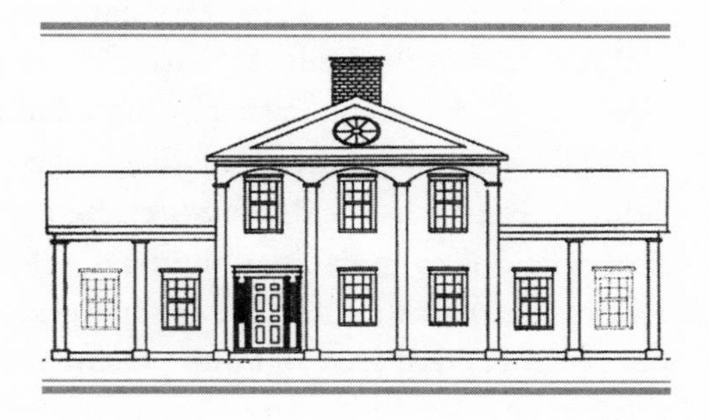

The Secondary Imagination

It's a miracle that curiosity survives formal education.
—Albert Einstein

In the months before Leo and Mark appeared, I had struggled to imagine the look of our house. We could have hired an architect, but a skilled professional would have cost money and our funds were limited. I must admit my ego factors in here, too. I wanted to be able to say to later visitors, *We designed and built this*. I was ready to take the credit and the blame.

Betsy and I were setting out to build a house that had a historic feel. The simple fact is that few architects understand traditional American buildings. The balance has begun to shift, and in recent years, a growing number of architects have opened their minds to old styles. Still, most designers continue to build new, not old, and they condescend to the very notion of reproducing the antique, refusing to repeat the past (in the memorable words of one influential modernist architect, "Imitation is suicide"). So we decided to go it on our own.

Well, not quite on our own. I believe that somewhere in everyone's brain there is a sort of slide tray that's packed with pictures of places. Most of these mental carousels have images in common, like the Eiffel Tower, the Empire State Building, and Big Ben. Many of these mind pictures are individualized, like the view of the house I lived in as a kid, Betsy's elementary school, and a couple of religious or municipal buildings. There are interiors as well as exteriors, close-ups and landscapes, and some foggy, indistinct pictures that we can't quite identify.

Some of these images—architectural memories, I like to call them—came to my mind and to Betsy's in thinking and talking about our house.

My childhood was spent in a onetime mill town in central Massachusetts. Like Red Rock, my hometown had been a prosperous place in the first half of the nineteenth century. Farms had flourished despite the rocky soils, and the uneven terrain produced streams that powered newly invented machinery. The fortunes of the town had dimmed when the railroad bypassed the village in the 1840s, and in the decades that followed, many farmers moved west, lured by the promise of rich, flat farmlands.

If I close my eyes and think back to the streetscapes of my childhood, the vision is of wooden nineteenth-century houses decorated with classically inspired details. Just below the roof overhangs there are broad horizontal bands (the architrave and frieze) and, on the corners, the flattened columns (pilasters). The Industrial Revolution had democratized architecture, making planed boards and machine-made architectural details affordable to the middle class. The added elements endowed these otherwise plain buildings with a formality and a sense of history. The simple boxes with gable roofs all looked distinctly like classical temples, whether they were churches, civic buildings, or houses.

While I tend to think in broader strokes, Betsy's focus is more likely to be on finer details. She had begun collecting nineteenth-century antiques a good many years earlier. She inherited the enthusiasm from her mother; their great affection for each other seemed to nourish their passion for antiques. After we were married and began our life together in Red Rock, the objects Betsy and I acquired suited our limited resources, coming as they did from local estate auctions. That also meant we often knew specifically where individual pieces had come from.

To us, the prescription was clear: we needed a home that was of a piece with the houses that survived in Red Rock, that would suit the furniture and objects we had collected, and that felt familiar and would please our eyes. The challenge was to transform years of stored images into a plan for one house that suited our family's needs. Over the years, we had talked about different shapes and forms. Betsy had dismissed the idea of a Cape Cod house ("The layout is boring"), and I had agreed, realizing I wanted a house that was at least distinguishable if not distinguished.

That was about as specific as our thinking had become, but I had set aside a few weeks to do the drawings before Leo's scheduled arrival in early July and Mark's a couple of weeks later. According to my plan, Mark and I could get most of the framing done before he went back to university in September; then I could figure out a way to get the house closed in before the first snow arrived in late November or December. To accomplish that, we had to put our dream house on paper first. But where to begin?

As an adult, I have developed a few strategies that come in handy when I find myself in a hard place. One of those is misdirection. By deliberately dancing around the perimeter of a predicament, I often stumble over what I'm looking for.

When something is missing, whether it's on my desk or in my workshop, I quit looking for it after a few minutes. As my patience grows thin and my anxiety rises, I move on, specifically to cleaning up the mess that caused the object in question to disappear in the first place. Most things have assigned places, so this goes back here, that down there, and so on. Before the clutter gives way to genuine organization, the missing item almost always emerges.

Operating under the misdirection-is-sometimes-the-best-approach theory, I decided to begin designing my house by learning how to design on a computer. After asking around a bit and pricing the options, I bought a software package that would enable me to do the drawings on my screen, using a mouse rather than a pencil.

Drafting software has become so ubiquitous in the architectural profession that it seemed to me that computer drafting *had* to save time and offer some other advantages. Many drawings for a house are alike; for example, the floor plan, foundation plan, and some of the framing plans share many of the same elements. That means that drawing each of them individually on a board involves redrawing the basic shape, the openings, and lots of details. I figured time could be saved by drawing one template on the computer. Then that file could simply be called up, renamed, and amended for various applications.

I bought a program that was billed as the amateur's equivalent of the computer-aided design and drafting (CAD) programs that architects use, and set to work to learn the software and to explore my house notions in a schematic but visual way.

For some people computers are good fun. For me they're tools, not toys, which means that assimilating new software is like making a new business contact. The introduction is formal (you don't know each other, after all) and the first exchanges are vague, perhaps even clumsy. But soon you find some common ground; you recognize something that breeds a certain familiarity. The program may or may not become your best friend, but soon you feel acquainted with it.

There were the usual setbacks—computer crashes, lost work, and silicon culs-de-sac that required calls to technical support. But within a week the computer was aiding progress rather than hindering it. I felt as if I were circling, readying for landing.

ODDLY, IT WAS A poet who, in a way, charted our course. Before Language Arts superseded English as a course of study, the literature syllabus consisted almost entirely of writers who were undeniably dead, white, and male. Samuel Taylor Coleridge was among them, and on our way to high school degrees, most of us encountered his poem "Kubla Khan." It was acknowledged by some of my adolescent peers as pretty good, even if it was by a dead guy. (At that stage the knowledge that old was outré was instinctive; later came the news that male and white were, too.)

"Kubla Khan" was different from other poems on the required reading list in part because of the circumstance of its composition. Coleridge claimed to have fallen into a deep sleep after taking two grains of opium. He slept for several hours and, upon awakening, took pen in hand and furiously transcribed the lines of the poem as if from memory. In the British Library in London there's a fair copy of the poem in Coleridge's own hand. It's undated, and the sheet has a number of corrections and scratch-outs. The writing pitches across the page; it looks to have been written in a hurry, as if Coleridge was in a race to put the words on paper before the clarity of a reverie was lost. When I was a student, the document—and the story—put a little drama into literary history.

Dreaming up poems and drawing up buildings are far from the same thing, and I wasn't about to buy some street opium to test its creative efficacy. On the other hand, I thought Coleridge was on to something when he talked of what he called the "secondary imagination." For him, the creative process required that the human mind take its accumulated perceptions and then manipulate them. In his words, the secondary imagination ". . . dissolves, diffuses, dissipates, in order to recreate." According to Coleridge, the act of creation is, in essence, a process of "echoing" the images collected as primary perceptions in order to shape them into new creations.

The story of how "Kubla Kahn" was composed may seem a bit far-fetched. Still, imagining something new, anything different, surely involves taking the known and, as the man said, dissolving, dissipating, and diffusing it. I share Coleridge's belief that a set of perceptions collected in the cerebrum can, if you listen with care, send distinct echoes.

Sometimes, claimed Coleridge, the conception arrives full-blown, and architectural history is not without moments when its geniuses experienced such epiphanies. Perhaps the most famous of them concerns Frank Lloyd Wright's Fallingwater, one of the most memorable houses of all time. According to his adoring apprentices, Wright designed the house, on deadline, in a single morning. He had nothing on paper when the wealthy client called to announce he was making an unscheduled luncheon visit. So Wright sat down and designed the house, creating in a few hours drawings that ordinarily take many days or even weeks and more. With a flourish, he wrote the name Fallingwater across the last sheet.

Wright's furious sprint produced an unprecedented design. It was far from what the client had expected. The commission had been for a "waterfall cottage," and the understanding had been that Wright was to design a house to be situated on a nearby hilltop with a view of the watershed below. Instead, in a moment of inspiration Wright had set the house into the opposite hillside, much of its living space literally above the rushing torrent of water.

Perhaps the explosion of creativity that produced the drawings that morning was a product of the secondary imagination. How did Wright conceive Fallingwater? For once in his life Wright, notorious for telling tales to his advantage, was silent. He never had to exaggerate anything about Fallingwater.

Whether he would recognize the term or not, the contemporary architect Hugh Newell Jacobsen has been known to rely upon the secondary imagination. One house of his in particular serves as a case in point. The client was Philip Palmedo, a New York entrepreneur. Palmedo himself told me the story.

He owned a piece of waterfront property on the north shore of Long Island about an hour's drive east of Manhattan. He wanted to build a house there that took advantage of the view of the Long Island Sound and Connecticut on the opposite shore. He also wanted an impressive, top-of-the-market house that, on its own, made a strong design statement. His first step was to hire Jacobsen, a Washington, D.C.–based architect, some of whose domestic designs have been widely published and admired in the architectural press.

Jacobsen spent many hours with Palmedo and his wife. In a laborious series of meetings and discussions, Jacobsen strove to develop a very detailed sense of the Palmedos' needs and desires, which meant lots of issues got resolved before the drawing process began. Jacobsen could tailor the design because he was familiar with how the Palmedos lived in their existing home and had a solid understanding of their expectations for the new house.

Such preliminary exploration is flattering to the client, but I for one don't believe that was Jacobsen's primary motivation. I think he wanted to keep the tinkering with his design to a minimum. Under the guise of fine tuning, the raw beauty of a designer's initial inspiration can easily be lost. Jacobsen's reasoning, then, might have been, *Give them exactly the elements they want, and with luck, they won't want to fiddle with the packaging*. It's the architect-as-artist approach, and the message is, *Don't mess with my work*.

I'm putting words in his mouth, but whatever his rationale, Jacobsen took what he learned from the Palmedos and went off to design their house. Somewhat to the clients' surprise, the initial getting-to-know-you stage was followed by silence. Months passed without a word.

Phil Palmedo remembers wondering after a while whether they would ever hear from Jacobsen again. Then one day he was working at his desk when his secretary buzzed.

"There's a strange man on the phone for you," she told him. "He says he's 'given birth.' "

Palmedo knew immediately that the caller was Jacobsen.

Jacobsen told Palmedo that while he was traveling back from Paris, the design solution for the house had struck him. He hadn't been drafting madly like Wright, working to a deadline. He hadn't even been at his drafting table, sketching and musing. In fact, he had no proper drawing materials at hand when the broad outline of the scheme had simply come to him.

So, seated in the pressurized cabin of a jet some thirty thousand feet over the Atlantic, he had memorialized the idea on an Air France napkin.

After Palmedo told me the story, I asked him immediately if he had the napkin. He shook his head, obviously wishing he did. I was disappointed, too. Even for those of us who believe in the secondary imagination, encountering concrete evidence of it occurs all too rarely.

I WISH I COULD report that a Kubla Kahn–like vision appeared to me. But that didn't happen.

In the absence of a spontaneous design solution, we worked at establishing a program for the house. Betsy and I drew up our wish list of rooms simply by looking at the home we had been living in for ten years. For the most part, the cottage had suited our needs nicely, so the task was to identify what spaces we felt we needed in addition to those we already had.

Sarah and Elizabeth were to have their own rooms, but we wanted at least one extra bedroom. We often entertain at sit-down dinners, so a dining room would be necessary in the new house, too. The living room, we decided, would be a public space, a parlor in the traditional sense, while a family room would serve more private needs and, if we were lucky, might also keep the toys from infiltrating the rest of the house. Television is not an everyday factor in our lives, so the TV could go in the family room, too.

The list didn't quite make itself, but without much difficulty we began to itemize the essentials. A library space would house the countless books that accumulate around us. I needed a room for use as my office. In the interests of economy, we decided to fulfill those compatible needs with one space for both library and office.

One of Betsy's enthusiasms is food. She has written books about entertaining, catering, and other food subjects. Whether it's just family or we have guests, the kitchen is always a focus of energy in our home. The kitchen in our new house, then, would need to be a space large enough to accommodate not only the work of food preparation but also eating and visiting.

We live in a cold and sometimes muddy and snowy place, so we would need a mudroom for boots and snow attire.

We both agreed that the main public space, the living room, was to be at the front of the house. Upon arrival, guests would then be able to admire the front façade, which we thought should make a statement. The bring-in-the-groceries entrance for family business would be at the rear.

The dining room had to have good afternoon light, as a setting sun at the dinner table casts the right glow. So it would be on the west side of the house. As a formal room, it could flank the living room, looking out the front and getting late afternoon light. Perfect, right?

"Well, no," Betsy observed, "the dining room really has to be adjacent to the kitchen."

"That's easy," I followed up. "We'll put the family room in the front, with the dining room behind, right next to the kitchen at the back of the house."

Betsy countered that we might need to think the matter through a bit more. The family room was to be the repository for the casual elements of our lives—loosely translated, that means a busy, messy room. I read her tone as *Not at the front of my house, you don't!* Our interests were essentially the same, but if you had witnessed this or another of our deliberations, you might have guessed we were adversaries.

Typically our discussions are self-contained, lasting a few minutes and, in the end, producing mutually acceptable answers. Sometimes, though, no answer is reached, and we postpone the discussion. My gloss on this approach is that it indicated our relationship has matured: in earlier years, recriminations would occasionally slip in (such as, "If you feel that way, there's no point in even discussing it!") and the discussions would end in small fits of temper. Occasionally, they still do. China doesn't get broken, mind you, and no household objects take flight, nothing like that. All in all, we have always had a healthy ability to discuss things in order to reach a satisfactory resolution, although our progress is, on occasion, rather slow.

In this instance, we needed to try another approach.

ANY GOOD HOUSE DESIGN must respect its site, and we looked to our acreage for clues.

A row of massive sugar maples lined the frontage of our property along Stonewall Road where Charlie Briggs hammered spouts and hung galvanized pails to collect sap during syruping season. Living nearby, we recognized the appearance of the pails as a harbinger of spring.

After we bought the land, I got into the habit of entering the property on the old cart road that angled between the two largest maples. Consisting of nothing more than two stony tracks on either side of a grassy hump, the path paralleled Stonewall Road about twenty feet into the property. Perhaps forty yards on, it turned deeper into the pine woods and meandered toward the center of the acreage. Eventually it curved back in a broad arc and rejoined the entrance road. The overall shape was that of a great lasso.

Within its loop was a knoll that was the highest point on the property and the obvious place for the house. The site didn't have a panoramic view, but what it did look out upon consisted exclusively of the acreage contained within the bounds of our land. The terrain rolled downhill from the knoll. It flattened out before rising steeply to a ridge at the far boundary. There a thick line of trees stood like the walls of a medieval town, a barrier between the generous clearing at our house site and the landscape beyond.

We could shape the open space as we wanted. Betsy was a practiced gardener and had even collaborated on a flower-gardening book a few years before. I'd grown a few vegetables and had always wanted to learn about trees, so landscaping the acres would provide us with the opportunity to plant ornamental and fruit trees. From the house itself, we could position the windows to frame views of our landscape. Even though the land didn't offer a commanding view of somebody else's mountains or lakes, it would, in a sense, allow us to control our own destiny, since no one could build within sight of our house.

But what was our home to look like? I thought we might take our cue from the primitive hut.

The primitive hut isn't a real building, but a number of theorists have employed it to explain the early development of architecture. One French writer of the mid–eighteenth century, the Abbé Marc-Antoine Laugier, saw the primitive hut as the precursor of all buildings. According to Laugier, the hut began with four trees as uprights. They supported horizontal wooden beams. Angled rafters of logs and branches were, in turn, attached atop the beams to frame the pitched roof. For Laugier, the hut was nature transformed, a pickup-sticks structure of plain posts and beams.

Other writers over the centuries have adopted the primitive hut as their own, but it was Sir William Chambers, an architect and surveyor-general to England's George III, who helped make the primitive hut the theoretical prototype for all classical buildings. Chambers transformed the uprights into columns, and his ceiling beams formed a frieze. The roof and rafter tails overhanging the beams became the cornice. The primitive hut, a rustic shelter loosely based on a daydream of Vitruvius, was suddenly recognizable as papa to the Parthenon.

No one will ever be able to prove that the theory is correct, of course, since the first wooden temple that Chambers and others believed was the father of all subsequent such buildings rotted to dust many centuries before the birth of Christ. Mythical though it may be, the logic of the primitive hut does ring true, and classical buildings speak to many of us in some elemental way. The columns hold up the broad spans; the effect conveys an elegant sense of function. The primitive hut provided basic shelter but did it with style. Surely that's why its descendants—of stone and wood, in Europe and America—are so numerous and the genus so revered.

Think of a postcard-perfect image of a little New England town. The view is from a nearby hillside, the village nestled into a valley, and there's a little white church at the center. Remove the steeple (that's a baroque elaboration), and the simple structure you are left with is a diminutive wood-frame version of a classical temple. Now, look more closely at the postcard and you'll probably see lots of other houses in that town that have the same temple front. I've been surrounded by buildings in that vernacular style for most of my life.

Some variation of that basic form was what we wanted to build, and one day, on our way to someplace else, we got a glimpse of a possible prototype for our future house. It was in a town in nearby Massachusetts, on a road that had been in continuous use since the eighteenth century. The houses that sat hard by its shoulders testify to its vintage: There were Georgians and Federals, Greeks and Victorians, and even an octagon. There were major houses and minor, handyman specials and virtual palaces. The place that caught my eye occupied a middle ground, both in scale and in vintage.

The narrow, deep shape of the house suited our site. The house had a gable front, with a long roofline pointing to the rear. It had a tall, triangular pediment. The place had probably been built in the second quarter of the nineteenth century as a house, but it had later become a library. We stopped to look more closely. The building was locked and shuttered that day, but I paced off the perimeter of the structure. The look and the size of the building seemed about right.

When I returned home, I began to put my new computer-drafting skills to work. In a couple of hours, a workmanlike sketch of the front elevation emerged, representing the two-dimensional appearance of a façade, a version of the remembered shapes of the house we had seen. It was little more than a quick compilation of geometric parts, with a slightly flattened isosceles triangle on top of a squarish, two-story base. There were six openings in the façade, with three windows on the second floor aligned above a door (on one side) and two more windows on the first. Lines of clapboards, pilasters, and trim added texture. It seemed to be working.

A rough floor plan came next, using the set of rectangular room spaces we wanted. Like a child trying to assemble a simple puzzle, I maneuvered the pieces—living room, dining room, kitchen, mudroom, downstairs half-bath, library-office, family room—into various positions. But each time a logical pattern seemed about to emerge, one or another element wouldn't quite fit.

I was frustrated, my arms folded across my chest, when Betsy came in to have a look.

"I like it," she offered, looking at the front elevation.

"Thanks. But the floor plan is refusing to cooperate. The pieces just don't seem to fit together." She looked over my shoulder as I fiddled some more. After a few minutes I was about to give up. "Any suggestions?"

She had none.

Why wouldn't all those rooms fit the form?

The overall shape could be enlarged, but we didn't want to build one of those monstrous houses that look all out of proportion. You know the kind—"McMansions," as some wag has dubbed them, the fifteen- or twenty-room monsters that look bloated and completely misplaced on their one-acre plots in expensive suburbs. That wasn't what we wanted at all: our notion was for fewer than ten rooms, and we hoped to fit them into roughly twenty-five hundred square feet. Our house was supposed to look like it belonged, geographically and chronologically.

We hadn't yet found the right solution.

FOR BETSY AND ME, the building of a new house had been a matter of occasional consideration for many years. As anyone does with a long-term plan, we had been stowing away our and other people's clever ideas.

A laundry room would be needed, of course; less obviously, we wanted it on the second floor, where the bedrooms were to be located. The cottage in which we were living had a cramped main floor and no cellar, so in remodeling it we had to locate the washer-dryer upstairs and came to believe immediately that it was a better way. Why carry all your laundry two flights down to the basement, wash it, and dry it, only to schlepp it back up two floors?

We also wanted to locate the laundry room close enough to the master bedroom so that we could have a pass-through linen closet. We had seen one in a friend's home. After you fold the towels, you open a cupboard door an arm's reach away from the dryer and put the warm, fluffy towels directly on their designated shelf. At a later time, upon stepping from the shower, you open the cupboard door on the other side of the closet (this one's in the master bath) and grab a clean towel. It's a small convenience, perhaps, but easy to design.

Much of the magic of a livable house is in such details as a central-vacuum system. Given that otherwise well domesticated American males tend to take a certain pride in how sensitive their ears are to the noise of the vacuum, doesn't it seem logical to put the motor in the cellar or garage, where it can disturb no one? The collection can goes with the motor, too, which means that when it comes time to empty it, you don't have to contaminate the living area of the house with a cloud of the dust and dirt that has just been laboriously removed.

Betsy did some homework. She found that a central-vac system is easy to install. In new construction, it's a simple matter of cutting and gluing plastic pipes together. She also learned that the machines aren't prohibitively expensive, since some manufacturers of portable vacs also make central-vacuum systems that are competitively priced with high-quality portable units. So a central vacuum was in the plan, too.

Good ideas come from all kinds of sources. One evening a friend regaled Betsy and me with stories of his sister-in-law. The lady in question lived in a wealthy New York suburb. In our friend's view, Lannie personified much of what's wasteful in our society. He thought she simply threw her money around. Lannie didn't work outside the home, yet she had a full-time housekeeper; she dressed her children in wildly overpriced Italian designer clothes, thought recycling a waste of *her* time, insisted upon expensive automobiles, and demanded of her husband nights at the opera—at which she invariably fell asleep and snored.

We only half shared our friend's outrage until he told us about her new kitchen. His sister-in-law hadn't been content with investing about $25,000 in mahogany cabinetry and another $15,000 in soapstone countertops, and all this in a none-too-grand house. She also had to have *two* dishwashers. What *could* she have been thinking?

Even before the coffee was ready the next morning, I raised the subject with Betsy. "You know that business of the dishwashers last night?"

She looked over at me quickly.

"It's not such a bad idea, is it?" I had thought this through at about three o'clock in the morning.

She cocked her head, smiling a little ruefully.

"I mean, in a new kitchen the cost wouldn't be much more than the price of the machine."

Betsy looked thoughtful, then reddened slightly. "If we do it, do you think our friends will snicker at us like we did at Lannie last night?"

"Well, we do entertain a lot. And you know how fast a dishwasher fills up when we put a dozen dinner plates in it and follow up with the dessert and salad plates. And the glasses. With no pans at all and . . ." I stopped when I realized Betsy was laughing.

"You don't have to defend it to me," she said. "I think it makes sense."

So we ceased finding the idea ridiculous. That morning, the mouse on my computer added a second dishwasher to the evolving plan.

LONG BEFORE WE TALKED to Charlie about purchasing property, I had pictures in my mind of possible house designs we might build. One of my favorites was a symmetrical three-part structure consisting of a central block with a gable front and two smaller, identical wings on either side with pitched roofs that ran perpendicular to the ridge of the main structure.

I really liked several nearby houses built in this configuration. They were trimmed out with pilasters and lots of details found in the buildings of classical antiquity, such as arches, urns, and swags. No one would have confused them with true Roman or Greek buildings—they were of wood, much smaller in scale, and looked like houses. But the classical inspiration was apparent.

When we first saw these houses, I understood only vaguely why they were described as *Palladian.* Later, I discovered they owed much to the work of Andrea Palladio, the sixteenth-century Italian architect. Palladio himself had studied and even measured ancient monuments; he had gone on to build many memorable villas in the Veneto, the rich agricultural region surrounding Venice and his native Padua. But it was his *Four Books,* a work that featured not only his prose but fine woodcuts of his remarkable designs, that made him arguably the most influential architect of all time.

I'd paged through *The Four Books*—a cheap reprint copy sat on my office shelves—though I couldn't pretend I had studied it with care. But one thing I knew instinctively on looking at local, vernacular three-part houses was that their symmetrical design required a level building site. An ideal location was at the crest of a small rise that, like a great earthen pedestal, would provide a suitable base for a three-part house.

Unfortunately our knoll in the blackberry patch was more like the broad back of a whale. The grade dropped precipitously away on one side; at the front it fell off steeply, then pitched and yawed. For those reasons, the Palladian idea had gotten filed away for another time and place. But several weeks into the process of trying to design the house, his tripartite plans came to mind. Having plugged away at different configurations and set them aside as unworkable, I thought a variation on that old theme seemed worth considering. It was another way to approach the problem, not necessarily a final destination but a tack to take. Worries about the landscape could get resolved later.

In a matter of hours, I was able to show Betsy a preliminary sketch of the front façade. She liked it. Next we needed a rough floor plan. Elements from the first drawing were incorporated into the next, so that the exterior openings on the first drawing, the windows and doors, appeared in equivalent places on the second. That provided the basics for the front wall; from there, I made an educated guess as to how deep the building might be, and planted a chimney in the middle.

A hall ran down one side of the main block, making that portion of the building conform to the well-established tradition of the sidehall colonial. The living room fit into the central block, with the kitchen behind. The master bedroom was upstairs, in the front; the girls' bedrooms were above the kitchen. The dining room, family room, and other downstairs spaces got assigned to the wings. The bathrooms, kitchen, and laundry room seemed to line themselves up atop one another in two stacks, simplifying the plumbing. The public areas divided themselves from the private areas. The floor plans took shape surprisingly quickly (see page 299).

As the lines shaped the spaces, our expectations colored what we saw, providing a larger context. An owner designing his own house anticipates moments in his own and his family's life. Betsy and I wanted a surrounding landscape in which our children could experience things, from the flower garden to the woodlot, from the frogs in our theoretical pond to the uninvited deer in the vegetable garden. We wanted Sarah and Elizabeth to understand something about the climate firsthand, from feeding wood into the stove in winter to keeping their windows open to catch cross breezes in the summer.

Another ambition of mine was to build a house with a yard and a façade of sufficient size and grandeur that my daughters would want to be married there. If anyone had confided in me a few years ago that he harbored such a desire, I would have classified that person as peculiar. Yet such thoughts did come to mind as I worked on the drawings.

THE FRONT ELEVATION HAD been settled, and Betsy and I had resolved the arrangement of rooms. She got the kitchen where she wanted it, and the dining room went where I felt it had to go. But the second floor was a problem.

The bedrooms for Sarah and Elizabeth were simply too small. I had tried rearranging the rooms, but the stairway and halls anchored one side of the house. The chimney mass, its position established by the living room and kitchen spaces below, sat squarely across the center of the house, immutable as only masonry can be.

Betsy came and looked over my shoulder one evening after dinner. She could sense the strain. Time was growing short, the process growing tiresome, and we still lacked some of the basics.

"They're too small," I told her. "Both of those bedrooms are just too small."

She looked again at the drawings.

"I just need about three feet," I told her. "The rooms are wide enough but too short."

She moved to leave when we heard Sarah calling from the other room. As she left, Betsy suggested over her shoulder, "Why don't you make the building longer?"

In an instant, my temperature rose about five degrees. *Because the tax man is already going to have his way with us,* I said to myself angrily. *We can't make it any bigger because we can't afford to, that's why.* I was almost snorting, my frustration at a boil.

Yet it wasn't many minutes later that understanding dawned: Betsy's solution was the right one. Less than an hour after that, the new floor plan of the main building was three feet longer and the bedrooms were large enough. The last of the basic decisions had been made, so I could go about completing the plans, and in the next few days, the final drawings emerged surprisingly quickly.

LATER, WHEN DESIGNING HAD given way to building, I often found myself at a loss for how to do something. At such moments in the construction process, I'd return to my little library of building references. One afternoon not so many weeks into construction, I left Mark at work and went to confirm how we ought to complete a framing detail. I pulled a couple of volumes off a shelf and, behind the books, noticed a long cardboard tube. I didn't know what it was, so I pulled it out to look.

The return address on the mailing label cited the Library of Congress, but that still didn't identify the contents. A look into the end of the tube revealed tightly rolled papers. They were architectural plans, elevations, floor plans, and details. The border of each drawing bore the name of the Historic American Buildings Survey.

I realized then what they were, but it took me a few minutes to realize what they represented.

Five years earlier, in an out-of-print book from the Depression era, I had encountered photos and drawings of an 1820s house in Canastota, New York. The Federal-style home had been recorded by the Historic American Buildings Survey. Originally an element of President Franklin Delano Roosevelt's elaborate reemployment system, HABS had for decades put architects to work recording important buildings. The result was an incomparable archive, which is today administered by the Library of Congress. Over time, I had ordered materials about several buildings, including the Nathan Roberts House in Canastota, which I had thought might later suit our needs. That was before we bought Charlie's blackberry patch. After perusing the images of the Roberts House, I had stowed them away for later reference and forgotten they existed.

As I looked more carefully at the Roberts House drawings, it dawned on me that there, in the unmistakable blue hues of architectural photocopies, was the façade of my design. The shape and volume of the two houses and the arrangement of openings were remarkably similar. Our house would be two feet narrower; the wings on the Roberts House weren't quite as tall, and the floor plans differed substantially. Despite many small differences, however, the overall conception and even many of the details were unmistakably the same. Without consciously selecting it, I had adopted the Nathan Roberts House as the precedent for our home.

My memory is better for shapes and images than it is for, say, memorizing lines of poetry, but there is another reason why our house would resemble the Roberts House. There is an underlying logic to designs that have remained popular. The proportions of our home can be traced back not only to the Roberts House but to other houses in my own area as well. And there are discernible links to that early publicist for classicism in America, Thomas Jefferson, as well as to Andrea Palladio in Renaissance Italy, to the Romans some fifteen hundred years earlier, to the Greeks before them, and to the "primitive hut."

I hadn't consciously *copied* any one house, yet I had drawn upon some clearly imprinted recollections of those HABS drawings, using what Coleridge had called the secondary imagination. Along the time line of building history, related buildings keep appearing and reappearing, evidence of a process of transmission that brings elements of the past to the present.

The discovery was not a disappointment. We weren't attempting to invent something truly new and original; our goal had been to find a comfortable spot amid our architectural memories. By dint of much planning and a small mnemonic accident, we would be sharing a niche in a grand tradition that had been many, many years in the making.

The Matrix Materializes

The hands are the instruments of human intelligence.
—Maria Montessori

A week had been required to frame the platform. Standing atop the plywood surface, one had a paradoxical sense of being both at the center of the world and on a launching pad—either you had made it, or you were just starting off. Sticking out of a rectangular opening was the top of a rickety old ladder. It was a reminder of which one our reality was.

Mark was very game and a quick learner. He was wary of power saws at first, but they soon ceased to intimidate him. In an equally brief time, he was confidently spouting measurements in sixteenths. Some inbred sense of caution led me to wear safety glasses and ear protection, and he did the same. We proved an excellent team.

The next step would be the walls of the first floor. We were now above grade, working on top of the foundation. We could no longer step quickly into the shade of the cool concrete walls to shield us from the August sun; they were below us, a ladder-climb away. These were hot summer days, and Mark and I were exposed to the sun virtually all day, every day, on a work site that had little shade.

As blue-eyed, pale-skinned types, we had been told often about the dangers of sun exposure. But there was work to be done before Mark went back to university. We had only about a month until he was to fly to Edinburgh, so we couldn't just settle into a shady spot in the nearby woods and wait for the heat to pass. We devised several strategies.

The first was to start work early. Mark and I would meet in the kitchen for cereal and coffee. We would tiptoe quietly so that Betsy could turn over and go back to sleep; if Elizabeth awakened, Betsy would nurse her quietly, then spend a few extra guilt-free minutes in bed. Sarah, the early bird in our house, invariably came down to join us, usually for a frozen waffle and juice. Before seven, Mark and I left the family behind and went to work.

We would work all morning, with a brief break for coffee and a snack. Back at the cottage, lunchtime would be a respite of an hour or more during the hottest part of the day. I could do paperwork, order materials by phone, and plan the next steps in the process. Mark read the paper, did various jobs out of the direct sun, or played with Sarah.

When we went back to work, we often felt as if we were working in the baking heat of a desert. So we dressed like desert people. There were no burnooses at hand, but we found suitable substitutes. The solution was as follows: Take one large white handkerchief, and spread it over the top of the head. Position the front corner at a point roughly four inches above the bridge of the nose. Pull the opposite corner so that the white expanse extends evenly over the top and back of the head. If the handkerchief is large enough, there will be sufficient fabric to cover the back and sides of the neck. Next, put a cap on top—a baseball cap is best. Mark's choice was a Sesame Street cap that Sarah loaned him, with Big Bird and Ernie looking over its brim.

With our billed caps and kerchiefs, we looked a bit like escapees from *The English Patient*. We each wore sunglasses, too, with straps of psychedelic yellow that looped around the backs of our necks to prevent the sunglasses from falling to the ground. We regularly lathered on generous quantities of sunblock. It had to be applied carefully because sweat would periodically drip into our eyes, and any sunblock that came with it would produce a blinding, burning sensation.

We were disciplined in our work, but we took something of an ad hoc approach to the process. To my way of thinking, building or assembling anything is to be done by instinct. Since boyhood, I've always found that if making something means I must slavishly follow step-by-step instructions, my concentration quickly lapses. I mean, how much can you learn if you're operating within someone else's tightly circumscribed universe?

Not that good plans aren't valuable—they are essential. As we built, we often used the drawings I had made, referring to the short pile of letter-sized sheets at least ten times a day. They were on 8½-by-11-inch paper because that was the largest sheet my computer printer could handle. That meant our plans were much less detailed than typical blueprints, which are as large as the two-page spread of a tabloid newspaper. There were just enough details on the plans to meet what was required by local ordinances. That was a good strategy, advised a builder friend. The fewer the plans, he remarked, the fewer the questions the building inspector will ask. "You want to build your house," my contractor friend asked, "or do you want him telling you how to do it?" This suited me because the approach left ample opportunity for the fun of figuring out how to do things as we went along.

Since state building regulations require the stamp of a licensed architect or engineer, I had hired a local architect to review our plans. He asked for a few more drawings, which brought the small pile up to about a dozen pages, and recommended a number of technical changes. He suggested increasing the size of some of the framing lumber and recommended the use of some extra steel fasteners called hurricane clips to connect the roof rafters more firmly to the side walls of the house. After all, this was shortly after several studies had been published about the structural weaknesses exploited by Hugo, the hurricane that blew off countless roofs on the Florida coast.

As Mark and I proceeded with the framing, the plans were always available. We kept them handy on a clipboard hanging on a nail near where we were working. They represented the work to be done in the way that a map of a strange country tells you of its terrain. Yet no plans (or maps) offer all the answers.

In the case of an architect-designed house, the builder has someone with whom to share problems that arise for which the plans offer no clear solutions. The contractor can call the architect for an answer *(So, what are we supposed to do when we get to that, you know, the corner beyond the fire wall where the duct comes in . . .)*. Or, more likely, the carpenter can blame the architect *(He should have anticipated this; he just wasn't doing his job!)*. The builder can then move on to cursing the architect's incompetence *(Damn it, I hate architects. I knew that guy in his bow tie didn't know squat about framing)*. Finally, inevitably, the workers on-site will figure out how to solve the problem. In our case, I just cut out the middleman. Having no one else to blame, we would find ways out of tough spots ourselves.

In the framing process, there was a lot of translating to do from the plans. The elevation drawing for the front wall of the house, for example, specified the distance from the exterior corner to the first window. The dimension was exactly three feet. Another dimension specified how far up from the platform the bottom of the window was set—that one was two feet six inches. Combined with the specified overall dimension of the window (or the rough opening, the void in the wall required to insert the window unit), we had the basics. But that's as far as the plans went.

After studying the plans, we would lay paired lengths of two-by-six that would eventually be the top and bottom of the walls, called the sole and top plates. My job was then to figure out exactly where the vertical pieces would go. That's called doing the layout, and it involves marking the junction points at which the vertical members or studs meet the sole and top plates.

In a wall with no openings, it was a simple matter of drawing pencil lines across the plates to guide the placement of the studs at sixteen-inch intervals. That way the sheets of plywood sheathing, which would cover the outside of the frame, would begin and end where the studs were—three sixteen-inch bays equal a four-foot width of plywood, and six bays, the eight-foot length. That's why a four-foot module is basic to building with today's standardized materials.

The layout process got more complicated when window and door openings had to be figured in. Each opening had to be in a precise location, especially in a house like ours, where symmetry was the watchword. Rarely did the openings fall neatly so that the standard sixteen-inch intervals flanked them. The result was that the basic pattern of lines drawn across the sole and top plates for the full-length studs was accompanied by extra lines that indicated where the cripple and trimmer studs that frame the windows and doors were to be located. A marked-up plate ends up looking a bit like an oversize bar code.

While I figured out the layout, Mark cut the studs at a workbench cobbled together out of scraps of two-by stock. It was ten feet long, with a well at one end built specifically to hold a chop box. That's a power tool that consists of a small working surface with a motorized circular saw mounted at its rear on an arm that hinges. When the spinning ten-inch blade is brought down in a chopping motion, it cuts through a workpiece; when pressure on the arm is relaxed, a strong spring returns the saw to its original position, leaving it poised to cut again.

With the chop box set into the well, the entire workbench became a long tabletop for precisely positioning stock to make neat, square cuts. We measured the desired distance from the blade along the length of the bench and then clamped a piece of scrap to the bench to act as a stop. A fresh length of two-by-six could then be positioned, one end butted to the stop, the other square to the blade. We would cut the first piece, measure it to confirm it was the correct length, then cut more—lots more, all of them of identical length. Then we did the same for the cripple studs that would be above and below the windows and over the doors, as well as the trimmers that would be inside the window and door openings.

When Mark had cut a small stack of studs, he would bring them over to where I was laying out the walls and position them to coincide with the pencil lines on the plates. When the timing was just right, he finished cutting and moving the studs about the time I finished my marking. Then we could nail the wall together.

Working on the first-floor platform, we would lay all the

studs and the sole and top plates on their sides, positioning them with respect to one another as they would be in the wall we were assembling. Then Mark would hold the top plate while I nailed the sole plate to the bottoms of the studs.

An indispensable aide was a small air compressor that powered a stick nailer via a fifty-foot rubber hose. It was a last-minute purchase and a tool Mark took to referring to as Mr. Nail Gun. He chose the name out of respect for how much time and trouble the tool saved us.

The stick nailer weighed about fifteen pounds. It didn't really resemble either a gun or the tool for which it substituted, a hammer. The look was more like a giant stapler as reinvented by some fifties futurist. Inside the stick nailer's cylindrical head was a piston. When triggered, the mechanism drove a nail through the nose of the device at tremendous force. There are nailers sold for a range of purposes, from roofing and finishing to framing, but the nail gun we used easily drove sixteen-penny common nails. It would bury a three-and-a-half-inch-long nail into solid wood with a single *ka-thunk*.

The nail gun saved time. Even a skilled carpenter requires roughly five seconds to position a traditional nail where it is to be driven, tap it once or twice so its tip penetrates the wood, and then drive its full length into the workpiece with a few powerful strokes. In the same length of time, a nail gun can drive five nails. In short order, I could drive all the nails into the sole plate while Mark held the opposite end of the studs, then hand him the gun so he could nail the top plate.

A power nailer has another advantage. For my aging body, the gun was insurance. Driving great nails with a traditional hammer sends a shock along the length of the handle to the arm. My worries about my sore wrist were still with me, and I knew full well that a dull ache at night and the numbness in the ring and little fingers have been problems endemic among blacksmiths and carpenters for generations. Thankfully the pain was no more than a memory a couple of weeks into the framing.

BOTH BETSY AND I belong to that subspecies of humankind classifiable as pack rats. There were programs in my attic for virtually every play, concert, and performance I had ever attended. Many people throw away a newspaper at the end of the day if it remains unread. I was unable to learn that skill. After completing a sewing project, some people toss scraps of fabric into the wastebasket. Betsy's refusal to do so came to be represented by large bins of leftover materials. Our parents were collectors, too —of stamps, glassware, books, *Life* magazines, and all manner of other stuff.

Over time, my habit was to read my backlog of accumulated newspapers, and Betsy found uses for a surprising number of her remnants. In fact, our instinct to amass objects of interest and potential use served us well in building our new house. Over a period of several years prior to beginning our construction project, we had been accumulating a wide range of bits and pieces that we thought might be used in building what we had taken to calling our "new-old house." Our sources of goods were varied—antiques fairs, friends, and neighbors. Included in our collection of stuff were nails from Wilho Aalto.

Wilho had been the custodian at my elementary school. His "office" was the boiler room, and as a kid I felt it was something of an honor to summon him to help with a classroom emergency like a broken window or a leaky radiator. I couldn't help looking up, taking in the great cavernous space quite unlike any other in the school. The boiler itself seemed as big as a house, and that warm, dimly lit room was somehow not of the school but of another place. It was a little bit dirty, the residue of dusty coal and thick heating oil still about. Work was done there, not learning, and for some reason I liked the place.

When I was a student at Wilho's school, much of the inventory in a typical hardware store had been sitting on shelves and hooks gathering dust for years. The floors weren't likely to be shiny, but were scratched and scarred. Things weren't so much on display as stacked and stowed, often encroaching on the aisles. Retail thinking has rocketed past such outmoded notions, and today the emphasis is on clean and neat and bright, and on how many times the inventory in a square foot of retail space "turns" in a month or a year. No doubt the gain in profitability compensates the shop owners for the loss of atmosphere, but it's not the same. Gone is the sense that things are about to get done.

A hardware store I remember from my childhood closed its doors a few years ago. On a visit to my parents' home, I asked my mother about it, and she told me the business had moved from its old cramped quarters at the bottom of a brick Victorian commercial building to a more modern retail space down the street. There, no doubt, it could better display the housewares it needed to sell to survive. Quite by chance, she also happened to tell me that all the old nails from the store had been given to a local American Legion post, the very one in which Wilho Aalto was an active member. For me, the *old* my mother had added to the word *nails* was an intriguing grace note.

You see, it can matter a great deal whether a nail is old. In the 1890s, the wire nail became the norm in the building business. What we think of today as nails—spikes, finishing nails, and the rest—are made from wire. To fabricate them, a machine unwinds wire from a heavy coil, chops each nail to length, then shapes a head and point at the opposite ends of the shank.

Early in the nineteenth century, the process and the raw material were different. A man operated a machine that sliced individual nails from iron plate. They were cut off with a tool that resembled a paper cutter, producing a nail with a square shank that tapered slightly on two sides. It was a material-intensive and labor-intensive process, which is a key reason why the cut nail was later superseded in everyday use by the wire nail. But the cut nail had some advantages. For one thing, it was less likely to cause splits.

When in use as a fastener, the wire nail is easily distinguished from a cut nail: the wire nail has a round head, while the cut nail's head is rectangular. Today's finishing nails are "set," driven slightly below the surface of wooden trim. The resulting hole is then filled with putty. When a cut nail is used in finish work, it is simply hammered flush. The square head nestles nicely into the surface of the wood but remains visible to the discerning eye. Since I was planning to build a house in the spirit of the nineteenth century, I wanted the nail heads to be visible; I had also used cut nails in other building projects and I liked the results.

Given the vintage of the store, I thought some of the nails Wilho and his fellow legionnaires had been given might be cut nails, so I telephoned him. I was pleased he remembered me, though I wouldn't have taken it personally if he had not. In his thirty-something years at the elementary school, he must have seen thousands of children come and go. And I had been an unremarkable child. To judge by my elementary school photos, my most distinguishing characteristics were a sober expression and a cowlick. But I think he actually remembered me.

I asked him about the store. He told me that when it was moved to its new quarters, the nail display had been left behind. It was no more than a great mound of old wooden bins, broken boxes, galvanized trays, and nail kegs, but countless nails rested there. The owner had realized that sorting, packing, labeling, and moving tons of nails would be more trouble than it was worth. He'd called the American Legion and said, more or less, "Get 'em outta here and they're yours." Wilho and his fellow vets had gladly taken them away.

"What kind of nails have you got?" I asked Wilho.

"Most of them are gone," he reported. Quantities had been sold to local contractors, some to legionnaires.

"What's left?" I asked, a glimmer of hope remaining.

"Mostly old nails," he replied, sounding a little sad. Maybe he remembered me as a little boy who took setbacks rather hard.

"Tell me about the old nails."

"Well, you know, they're those square-shafted ones—cut nails, they call them. You know the kind. Not good for much. Kind of specialized."

I resisted the temptation to utter a long *Yesssssssssssssss!*

"Are there many?"

"I'm not sure," Wilho said. "A few boxes and a keg. Maybe three hundred pounds?"

He told me the price was fifty cents a pound. He said it a little apologetically, but I thought that seemed more than fair, knowing that although reproduction cut nails were available, some of them cost more than five dollars a pound.

I bought all the cut nails Wilho had on offer. That made him happy—until my call, he was afraid that his garage was destined to be the permanent resting place for five fifty-pound boxes and the single one-hundred-pound wooden keg. And it made me happy, too, and not because I saved a few dollars. More important, those weren't reproduction nails. They were "new-old stock." That is, they were unused but also of a certain age, having been made as much as a hundred years earlier. I like the idea of loading up a nail pouch with nails another carpenter seventy-five or a hundred years ago might have used.

When I went to pick up my purchases, Wilho was the same quiet but welcoming man I remembered. We loaded the nails into my minivan, then reminisced awhile about the town, the school, and acquaintances we had in common. Perhaps it was our agreeable conversation—or was it his relief at being rid of the nails?—but he gave me a gift. It was a screw-top quart bottle filled with a pale orange fluid, his own homemade tomato wine.

ONE OF MARK'S BEST attributes as a laborer was his dislike of idle time.

Many people promptly do what they are asked, cheerfully and well. But when they're done, their internal metronome seems to click to a stop. That never happened with Mark. Like a cat hunting for game, he would instinctively look around for something else to do.

At the start of the day, he didn't wait for instructions; he'd take the tarp off the tools and materials. When more lumber was needed at the saw table, he'd go and get the stock and then pile it within reach of where it was to be used. When the gun needed a new clip of nails, he'd get two and keep the spare in one of the pouches in his tool belt. Mark was born with an ability to anticipate, a gift not everyone has.

As we began raising walls, a pattern of movement quickly developed. Each wall had to be square, meaning its corners were true right angles. The tools called squares aren't much help—none we had was large enough to square a wall that was twenty-five feet long and eight feet tall. A tape measure, however, would do the job with ease.

Upon finishing the nailing of the top plate to the wall, I

would put Mr. Nail Gun and the hose that trailed him off to one side. Meanwhile Mark clipped the end of a fifty-foot tape over one corner of the wall and walked diagonally across the rectangular structure. Laying the tape across the far corner, he called out the length ("It's twenty-six foot six inches this way . . ."). By then I'd be at the other end where the tape was clipped over the corner, and we'd shift the tape to the other diagonal ("This one's shorter, more like twenty-six four and a half").

The dimensions usually were different, but there was a solution for shifting the wall into square. We would use a sledgehammer, though we liked to call it the persuader.

The first time we put it to use, I handed the persuader to Mark. He didn't know what to make of it, but I could see from his pleased expression that he had a young man's fondness for large tools. The persuader was bright orange and weighed almost twenty pounds.

"You use it like a croquet mallet, Mark," I explained. "A whack on the corner will shift the wall."

He understood immediately and positioned himself like a practiced croquet player at the corner. Then he took an arching backswing, preparing to give the wall a great crack. I didn't have time to call *Not too hard, Mark!* before there was a loud *clonk!* The wall jumped, the persuader having driven it a full foot, racking the wall much farther out of square.

Mark looked up, as surprised as I.

"Perhaps I struck with a little too much force?" he said quickly. Without waiting for a response, he moved to the opposite corner and tapped the wall back with gentle strokes.

Once we had confirmed a wall was square with another set of measurements, we would nail a couple of lengths of two-by stock diagonally across the frame to keep it that way. The wall was then ready to raise.

With a pair of small pieces of scrap (called scabs) nailed along the front of the building to prevent the foot of the wall from sliding off as we raised it, we lifted the top of the wall and walked toward the outside edge. A few grunts later, the wall was vertical, more or less. We would nail the foot in place, having aligned it with the outside edge of the platform. Then we plumbed the wall, getting it precisely vertical with a bubble level, and braced it in place with another couple of lengths of two-by. Then it was on to the next wall.

Framing has its macho aspect. It's big muscle work, requiring more strength and athleticism than, say, electrical, plumbing, or cabinet work. If a line of floor joists or wall studs varies a bit—say, one or two in a long row bow out an eighth of an inch—it won't matter much. All the framing members will disappear beneath the skin of the house, and minor variations will get averaged out, disappearing into the rank and file of surrounding members.

Yet the basic orientation must be right. You need more than plumb walls, as the platform has to be level, too, precisely horizontal. The platform and wall sections of the frame must be square to one another. Even minor problems with a wall that's out of plumb, a floor that isn't level, or a section that doesn't consist of right angles will get compounded as the house rises. Then everything that comes after—siding and trim on the outside, wallboard and casework on the inside—will have to be cut at odd angles. The compensation for one mistake makes the next stage more difficult. A compromise results, and one compromise leads to a series of others.

Framing is repetitive work. Oddly, that's one of its satisfactions. There are efficiencies to be devised as you go about doing tasks. One is that you learn to build walls as close as possible to where they will eventually stand. Moving a fully assembled wall is heavy work, so that's a mistake you quickly learn to avoid. You also learn to plan which wall goes up first, which second, third, and fourth. If you don't think the sequence through in advance, you can end up with an impossible raising job.

Framing offers the satisfaction that, after a long day's work, you have something to show for your labors. Even for a two-man crew like ours, several small walls went up on some days. Window and door openings were suddenly there to see and walk through. With a squint of the eye and a pinch of imagination, we could envision what was emerging.

We worked full days, seven days a week. That was Mark's choice as much as it was mine. He wanted to make money to supplement his college grant, and he found he enjoyed the work. After almost two weeks of nonstop framing, we did take a break. On a Saturday, Betsy and I wanted to go to an antiques fair at a nearby fairground. Mark and the girls came along, each of them about equally unaware of what they were in for. I carried Elizabeth, who had just graduated from a Snugli to a baby backpack.

More than a hundred dealers had set up tables and tarps and laid out objects for sale. A few sellers had spaces inside a barn-like building used annually for 4-H and other agricultural displays at the county fair, but most of the wares were outdoors under the sun.

We wandered around, looking for nothing in particular. We bumped into a few friends and neighbors and watched carefully to be sure Sarah didn't get too interested in displays of china or glassware. I introduced Mark to a couple of people, but he was preoccupied with taking it all in. A number of the dealers had architectural salvage, too, so I kept an eye out for pieces that might be put to use.

I spotted a pair of seven-foot-tall French doors. Each one was only about twenty-two inches wide, but they were intended to be hung in the same frame so that they would open together like the doors of a church. I liked what I saw and walked over to the dealer's booth to take a closer look.

The doors weren't very old, dating from the 1920s. Aside from a couple of missing panes of glass, they were in excellent condition. There were no frames, just doors, but the tops had flattened elliptical curves. They seemed like they might be a solution to a design problem we had.

Among the many unanswered questions in the building plans was where the point of exit from the dining room onto the deck would be. The deck would be on the west side of the house, and in a passing fancy, I saw these French doors as the answer. But we needed to fill *two* openings on that elevation, one from the dining room and one from the family room. To suit the design, the elevation had to be symmetrical.

I was about to walk away with a shrug and a sigh when the dealer approached me. We knew each other slightly, so we talked about the crowd (small) and the rate of sale (slow). I also told him of my dilemma. He laughed happily and told me he had another identical set back in his warehouse. "They would be perfect for that," he assured me. Betsy and I talked, together we examined the doors, and then I wrote a check for $695. After making an appointment to pick up the other pair, Mark and I roped the tall doors to the roof rack on my minivan. The day had proved productive.

On the way home, we stopped at the work site. The wooden structure appeared to be little more than an enormous set of boxes gradually rising out of the earth. Yet as the doors we had just purchased reminded us, we were building a labyrinthine matrix into which dozens of windows and doors, closets of all sorts, two and a half bathrooms, several tons of plaster, and a thousand other elements were to be fitted. A skin of clapboards, trim, shingles, paint, and decorative elements would cover the exterior.

Our plain boxes would become a house, with two entrances on the west elevation. Our purchase that afternoon had been well timed. Within a few days, Mark and I would find ourselves laying out the wall in which those doors would eventually be hung.

ONE REASON WHY AN amateur like me could presume to build a house is that house construction is primitive in nature. You begin with a pile of sticks. You stand some of them up. Then you grab some longer ones and lay them across the tops of the vertical sticks. The wooden frame that emerges, whatever its builder's skills, is common to virtually every American house.

Simple though the basic principles may be, labeling the result can get complicated. A wooden structure that fits this description is called post-and-beam, a name that refers to its vertical and horizontal members, respectively. But a structure of similar elements made of stone is termed post-and-lintel. There's also a fifty-cent word used to identify all such structures: they're trabeated, from the Latin word *trabes,* meaning beam.

Until the Romans devised the arch, virtually every structure with a roof was trabeated. The technique is ancient. The great megalithic monument at Stonehenge is a stone circle consisting of posts and lintels. The builders of the pyramids employed the same principles. Today a preschool child builds a trabeated structure by positioning two blocks with a space in between and laying a longer one across the top. It remains the single most important architectural contrivance and is the first rudimentary principle any builder learns.

In building my house, Mark and I used standard-dimension lumber, two-by-six-inch studs for the vertical members and two-by-twelve-inch horizontal joists. But until late in the nineteenth century, the posts and the beams in the American wood-frame house were usually timbers, lengths of wood that in section were at least six inches square. Some were much larger, often eight by ten inches or even twelve by fourteen. In thinking about our new-old house, we had considered building with timber posts and beams—there is a visceral appeal to the permanent and enduring look of a timber frame. Yet as we know from the stories of Goliath and Samson, powerful appearances can be deceiving. Engineering studies have found that stick-built buildings like ours, with their rows of smaller studs, are actually as sturdy as timber frames and under certain conditions even stronger. While a two-by-four may look like a matchstick next to a giant timber, walls and platforms built of carefully arranged two-by stock combined with a membrane of plywood are actually less likely to fail than more massive members made of individual tree trunks.

A manpower issue helped make the decision, too. A platoon of workers, probably a dozen men or more, would be required to raise a timber frame for a house the size of ours. I couldn't call upon that many volunteers nor had I the funds to hire such a crew. When it came to making the call, my heart told me to build a timber frame, but my head and my pocketbook told me that stick-built was the way to go. The two of us, Mark and I, would be able to do it.

MARK TRAVELED LIGHT—HE had arrived with only a duffel bag of clothes and a smaller bag slung over his shoulder. But along with a handful of books and personal items, he had found a place in that shoulder bag for a few precious CDs.

The CDs came in handy because one of the ways we amused ourselves during construction was with a boom box. Our musical tastes were rather different. I've been listening to Van Morrison since he recorded "Brown Eyed Girl"—that was about five years before Mark was born. Mark loved a range of English bands unfamiliar to me, such as the Stranglers and the Clash. We had the Beatles in common but my taste for female vocalists was at odds with his allegiance to loud, rhythmic guitar bands.

We listened to each other's music, all in a game attempt to be open-minded men of the world. In my CD collection I had a range of people he didn't know, from Pete Seeger ("Rather old, isn't he?" Mark commented) to Steely Dan (Mark would shudder at their smooth studio sound—"Are you sure they're alive?" he asked me once). I took to raising my voice as if I were talking to a deaf person when the Stranglers were playing.

At first we had taken turns changing discs when the boom box went silent. After a few days, I moved less often to select the music, and Mark filled in. A few days later, Mark tired of the task, and a sort of pantomime ensued.

We assumed the roles of the unhappy couple who, unwilling to acknowledge the friction in their relationship, express their irritation by passive aggression. I tested Mark's patience by ignoring the silence when the music stopped. Eventually, he climbed down from his ladder in exaggerated movements, accompanied by a dramatic exhalation or two, and put a disc on. Typically it would be a recording that was at the opposite end of the musical spectrum from my tastes. Usually I found his reaction amusing (though I kept my amusement to myself) and I would take the next turn, sensing his patience was wearing thin. The cycle would then be repeated.

After lunch one day, my turn came. Grudgingly but wordlessly, I climbed down the ladder from the second-floor platform to the shoe box of CDs next to the player. We had listened to most of them, but I wanted something different. I chose a recording of Aaron Copland's *Appalachian Spring*.

As the quiet orchestral strains of the piece began, I headed back to work. Mark was uncharacteristically still, standing at one corner of a sheet of plywood, waiting for me to join him in maneuvering it into position. I was wondering what he would think of the piece. It was a work-site experiment.

I didn't have to wait long for his reaction. I was bent over the plywood when he spoke.

"What *is* this?" he asked dryly. He waited half a beat before answering his own question. "Perhaps it's the cleaning disc?"

We both laughed, the kind of deep, surprised laughter that really has little to do with what apparently prompted it.

That day, and every day, Mark's energy led me to work as hard as I ever had, both to keep him busy and to keep up with him. I was enjoying the process of building, doubly so because of his intelligent and wry presence on the job.

ONE MORNING IN LATE August, the house took on the look of a Spanish galleon. We weren't hallucinating. The previous afternoon we had finished framing the first-floor boxes for the two wings, and they resembled the poop deck and forecastle of a ship. The structure seemed to float before us in the morning haze as Mark and I hiked along the driveway to go to work.

"Now, if we add a mast . . . ," Mark mused, half to himself.

I hadn't expected the house would assume the ghostly shape of a ship, but I had very much looked forward to seeing the balanced pair of boxes sitting atop the main platform. With them in place, the central block could begin its rise, too, and the three-part structure would start to look like a house. That morning we would begin positioning the second-floor joists between the two wings. But the way in which those joists were to be set in place wasn't usual. It was a structural solution that I had never seen but had devised to solve several problems.

Designing the building had required an understanding (if not mastery) of some of the legerdemain of the architect's art. Architects aren't magicians, but they are required to do one thing while appearing to do another: somehow they conjure up buildings out of mere sheets of paper.

Architects are the dramatists of the plastic arts. The works of poets, painters, and sculptors are experienced firsthand by the interested public more or less as the artists created them. In contrast, architects don't "make" buildings. Rather, they execute drawings (as dramatists write dialogue). Then the builders (or actors) take over, re-creating and, to varying degrees, interpreting for the public the architect's (or the playwright's) vision. I had had to learn the rudiments of the process that enables architects to communicate their vision at second hand.

Architects make orthographic drawings. The prefix *ortho* means straight, vertical, or right-angled; an orthographic projection (or drawing), therefore, represents an object as it would appear when seen straight on. We don't actually see things that way—the optics of the human eye and the way the brain interprets visual data are much more sophisticated. A perspective drawing more accurately simulates the eye's view, in which some elements in the drawing appear closer and some farther away. But perspective drawings are close to useless to the builder, since they are full of distortions.

The basic orthographic projections are elevations and floor plans. An exterior elevation describes the face of a building from the point of view of an observer looking from a horizontal vantage. A floor plan shows the location of the rooms in a structure, drawn as if from a helicopter hovering directly above a building that's had its roof removed. These highly disciplined drawings are festooned with measurements, and because they are done precisely to scale, they can be translated by carpenters into marching orders.

The three-part house to which we aspired had a clearly defined mass. Unlike the flat image in an orthographic drawing, a mass takes up space, like a book or a box. *Mass* is a word that's handiest in thinking about the exterior of the box, so my first step had been to sketch the elevation of the front of the house, reducing it to two dimensions.

Volume is the outside-in of mass. It's an indoor word, one that describes space, specifically interior space. While the exterior of a structure appears to be a solid mass, it actually encloses a three-dimensional space. That is, the volume. As Betsy and I had developed a scheme or program for the house, we gradually encoded our thinking into a series of working orthographic drawings that specified building details like lumber dimensions and rough openings. When Mark and I continued the process on the work site, we deciphered and interpreted the architectural renderings. The geometric language of the architect's craft is essential to each of these transmutations, from the blurry mind's-eye image of a mass, to a highly ordered set of flat drawings, to the interpretation of them into the full-scale, three-dimensional reality of a wood superstructure.

Challenges had presented themselves as I made the drawings. For example, we wanted tall ceilings in the main rooms, especially the living room and kitchen. Our goal wasn't to make a grandiose statement, but we wanted to avoid the sensation of a low ceiling hovering just overhead, which one gets in many early houses. I'm a bit more than six and a half feet tall, so I'm forever ducking through doorways. I wanted these rooms to welcome me. That's a matter of scale, another of those concepts, like mass and volume, that architects grapple with as they go about imagining full-size houses on small sheets of paper (or computer screens).

We also wanted to fit a full floor of living space upstairs. That was in conflict with our basic design because the volumes of the three buildings had to be different, with a tall box in the middle and noticeably shorter ones on either side. So I struggled with the problem of how to make the shorter wings tall enough to hold two floors of living space.

We worked with various dimensions, adjusting the proportions of the house, deciding on its width, height, and depth. We adjusted the shapes—for example, we made the wings as tall as we could to add ceiling height to their upper stories. But it didn't quite work. Even if we accepted the fact that the ceiling in those spaces would follow the pitch of the roof and angle downward to shorter walls at the front and back, the ceiling height at the center wasn't tall enough, only about six feet.

I had worried the problem for hours during the design stage. All we wanted were reasonably tall ceilings of nine feet downstairs and livable ceilings upstairs in the wings (seven feet seemed a minimum). But it just didn't seem possible to have both. There was another problem, too. The taller ceiling height that seemed appropriate to our large living room and kitchen spaces seemed out of proportion in the smaller rooms that flanked them (library, dining room, and so on). I began to feel like an arbitrator caught in the verbal cross fire between two uncompromising parties.

Then I came to the realization a trained architect would probably have come to much earlier: nowhere had it been decreed that the ceilings on the first floor had to be of uniform height.

The smaller scale of the first-floor rooms in the wings would be better served by lower ceilings. Not low—Betsy decreed eight feet was the minimum. But lower than in the living room and kitchen in the main block. By lowering the downstairs ceilings in the wings, we would gain essential height upstairs. In this insight was the germ of a solution.

Yet settling on the solution of two different ceiling heights didn't automatically provide a means of achieving it. A practical question remained: How would we frame the building with two different ceiling heights? There's the architectural idea . . . and then there's the carpentry implementation, the transformation of lines on paper to pieces of wood on-site.

The solution came to me some days later. It was surprisingly obvious, and its elegant simplicity pleased me. The answer was to build the two flanking wings first, our poop deck and forecastle. Then—here comes the ingeniously simple bit—the platform for the second story of the central block would rest literally on top of the floor of the wings, the ends of the floor joists overlapping onto the wing platforms. We're back to trabeation again, as a heavy horizontal sits atop sturdy verticals. In this case, however, the entire flanking structures were to become posts, carrying the weight of the main block down their sturdy bearing walls directly to the foundation. That was how I drew the house. (See framing plan, page 300.)

When we arrived at the galleon that hazy morning, I had begun telling Mark the story of the problem and the solution we would be putting into place. But his attention seemed to be wandering. I realized after a moment that, having glanced at the plan, he had absorbed what he needed to know. He didn't require an explanation of how I had stumbled onto the overlapping platform idea.

He just wanted to get to work.

Of Hearth and Home

Tradition is the means by which society maintains its identity in the face of change.
—Robert Adam

I built a chimney once. More accurately, I increased the height of an existing chimney by about ten feet. We were putting a second story on an ell of our cottage to add an upstairs bedroom. The top of the original chimney hadn't been in very good condition, so I had dismantled the collapsing courses of brick, taking careful note of how they had been laid up. I bought a reference book on masonry and concrete construction in a secondhand bookshop and asked a few questions at a masonry supply yard nearby.

That chimney top is still standing, but it's not an object of beauty. Getting mortar to the right consistency isn't as simple as it seems, so my runny mix oozed like pancake batter from between the joints, giving the bricks a grayish cast. The joints were thick and uneven. The task was time-consuming and risky, too, since the work had to be done on a makeshift staging twenty-five feet above the ground.

That chimney experience told us that this was one area where I had neither the skills nor the inclination to do the substantial masonry work required at our new house. Although the budget didn't allow for hiring many subcontractors, a search for a mason was necessary.

I asked around. The name of one local mason kept coming up, and his restoration work in several older homes had been first-rate. After leaving messages on his phone machine two or three times and receiving no return calls, I realized I might as well have been talking to myself. Another mason seemed interested, but when I asked him about Rumford fireplaces, his expression went blank. Count Rumford, I told him, was a contemporary of Ben Franklin who devised what many people still believe is the most efficient fireplace design. "You know," I went on, "the ones with the shallow fireboxes?"

His expression remained fogged with confusion. "No, I've never built one of those," he assured me. We shortly agreed that perhaps this wasn't a job for him.

Then a friend, Don, whose life is old buildings, gave me the name of Ralph Bruno Jr. I asked Don if he had ever worked with Ralph. He said no, but that meant little because, like me, Don preferred to do his own work, and unlike me, he was a first-rate mason.

"Have you seen his work?" I asked.

Again, the reply was no. A few questions later, it became clear that Ralph's enthusiasm had sold Don on him.

Don told the story of how they met. Ralph wandered up the dirt road to Don's home and found him building a stone wall. They talked. Ralph asked intelligent questions, and before long Don was pretty sure this solidly built stranger wearing wire-rimmed glasses had done a good deal of masonry work.

Then he learned why his visitor had appeared that morning: Ralph was out "calling," looking for converts. He was a man of faith, a Jehovah's Witness.

I cringed when Don told me Ralph was a Witness. While not begrudging anyone his or her faith, I would like others not to presume that I need to be told how to go about my worship. Don feels the same way, so when Ralph broached the subject of faith, Don bristled.

"You can put that right back in your pocket," Don said of the tract Ralph tried to press on him. "I'll talk masonry with you all day, but none of that religion stuff. Okay?"

Ralph went on his way, amicably, not long afterward.

He came back a few weeks later. He tried with the tract again, only to meet with the same rejection. He put it back in his pocket; then they talked about masonry, about the chimneys the Shakers had built in a community not far away, about the old methods of working with clay mortar and handmade bricks. In the way of accidental acquaintances with a common passion, they became friends. So Don gave me Ralph's name. "He knows what he's talking about," Don told me.

In the hundreds of conversations Ralph and I have had since, he never broached the subject of religion. On the other hand, he immediately brightened when, at our first meeting, the name Rumford was mentioned.

BETSY AND I KNEW that a fireplace wasn't necessary or practical; but we wanted one. We didn't need several—there's a vogue in new construction these days in which master bedrooms, dining rooms, kitchens, and even entrance halls have fireplaces. We thought one in the living room would be quite enough and, moreover, all we could afford.

Not just any fireplace would do. Given our desire to build a house that alluded to the past yet was also of the present, our fireplace would have to be appropriate to our house design. We wanted one that both looked the part and was highly efficient and safe. Count Rumford and the genus of fireplaces that bear his name offered the solution.

Despite his aristocratic title, Count Rumford (1753–1814) was born Benjamin Thompson just outside of Boston. At thirteen, he worked briefly as a clerk in a country store, then apprenticed to a physician. The same curiosity that would later produce improvements in fireplace design almost got him killed when he tried to replicate Ben Franklin's famous kite experiment. He wrote in his diary, "It had no other effect on me than a general weakness in my joints and limbs and a kind of listless feeling." Apparently the sight of him was more dramatic than his account would suggest, since family members reported seeing him silhouetted in fire.

A Royalist, Thompson expatriated in 1776 to England, narrowly escaping capture by a band of revolutionists intent upon tarring and feathering him. He became a full colonel in the British army before moving to Bavaria, where his star rose still further. He was made a general in the Bavarian army, later was minister of war, and was eventually given the title of count of the Holy Roman Empire. He styled himself Count Rumford after the original name of his wife's hometown of Concord, New Hampshire.

He kept a journal throughout his life, and in his notes he recorded wide-ranging experiments. An important series of trials that he conducted in Bavaria compared the thermal conductivity of various materials to determine which would make the most effective uniforms for soldiers. His findings led him to the speculation that it wasn't the material but the air trapped in the fibers that provided insulation, a notion that seemed bizarre to his contemporaries but today we recognize as patently true.

When he returned to England, London was choked with smoke. The ever-empirical Sir Benjamin (his knighthood compliments of George III) set about dealing with the city's air pollution. The problem began inside the houses of London, where the fireplaces were smoky, and he outlined his findings in a pamphlet, published in 1795, titled *Chimney Fireplaces with Proposals for Improving Them to Save Fuel, to Render Dwelling Houses More Comfortable and Salubrious, and Effectually to Prevent Chimneys from Smoking.*

I have an itch for discovery, especially of the bookish kind, and coming across the work of Rumford some years ago was a thrill. I had seen Rumford fireplaces in early American houses, since his prescription for the perfect fireplace had been widely adopted in the early nineteenth century. But as heating technology moved on, with stoves replacing fireplaces and, eventually, furnaces and boilers superseding stoves, his principles were somehow forgotten.

When I found his writings, there was a sense of having rediscovered one of the lost secrets of civilization. I remember thinking, *If only good design could always be reduced to simple formulas!* He offered arithmetical rules specifying, for example, that the depth of the firebox should equal the width of its back wall. Rumford decreed that the width of the front opening of the fireplace was to be three times the depth. Ditto for the height of the opening—it, too, should be three times the depth, according to the rules of Rumford. The sides were to be splayed at oblique angles so that the radiant heat would be reflected out into the room rather than contained within the firebox.

On first viewing, a Rumford fireplace is startlingly different from conventional, squat fireplace designs. The typical firebox is a deep, dark recess, but a Rumford smoke-reducing fireplace has a much taller, shallower firebox. Were you to bend and crane your neck in order to look upward from the firebox of a Rumford into the throat of the chimney where the smoke ascends into the flue, you would see a very narrow passage. In eighteenth-century England, the typical chimney throat was wide enough for a sweep to gain access to the flue. But Rumford insisted that it be reduced to no more than *four* inches.

Rumford's design modifications were counterintuitive. It seems illogical to push the fire toward the front of the fireplace and to shrink the opening through which the smoke is to rise in order to cut down the smoke that escapes into the living spaces. Yet these design alterations worked.

Rumford was, along with Franklin and Jefferson, one of the great polymaths of his time (Franklin made his contribution to heat technology, too, with his stove, although Jefferson adopted Rumford's design for his fireplaces at Monticello). All three men were, to use Franklin's term, "experimental philosophers" who pondered natural objects and phenomena for practical ends. Decades before the term was even coined, Rumford developed theories of thermodynamics. Until he came along, heat was thought to be a substance rather than the energy associated with the motion of molecules.

Not only did he reduce the smoky haze of eighteenth-century London, but he simultaneously increased fireplace efficiency. Recent tests conducted on fireplaces built to Rumford's specifications found that his radiate two to three times more heat than conventional fireplaces. They are also the only masonry fireplaces to meet the stringent emission standards that certain states have established. Rumford fireplaces smoke less, generate more heat, use less fuel, and burn cleaner. And, Betsy and I decided, they're historically appropriate to a house that might have been built in the first quarter of the nineteenth century.

MY FIRST MEETING WITH Ralph took place before the building of our new home had begun. The pickup truck seemed to be the standard mode of transport for all contractors, but Ralph arrived at our cottage in a battered American-made sedan. The car was a bit like Ralph: unassuming, sturdy, and practical.

Unlike the other tradesmen who had looked at them, Ralph studied the working drawings intently. He asked to have copies of several pages. He made admiring remarks about the look of the house and seemed to understand immediately what we were trying to do. I pointed out the living room fireplace on the floor plan, then showed him on the front elevation where the chimney was to emerge from the roof. While the living room wasn't centered on the front façade, symmetry demanded that the chimney be at the precise center of the roof. That meant the flue would somehow have to step over about two feet between the first-floor fireplace and the roofline.

Ralph started to nod as I explained. He looked up from the plan and interrupted softly: "We'll corbel it over." What he meant was that he could shift the structure, overhanging each course of brick half an inch or so. Within a few feet, the brick structure would then step over to the center of the house and could emerge from the roof where we wanted it.

Ralph's follow-up remark made me feel even more confident that he was the right candidate for the job.

"It wouldn't look right if it wasn't in the center," he concluded. He understood intuitively what we wanted.

When I told Ralph we wanted a Rumford fireplace, he looked surprised. "You know about Rumford?" he asked.

"I thought I was asking you the same thing."

He narrowed his eyes a little as he looked at me. I came to know that look and to think of it as Ralph's *I wonder if* expression. "It's surprising that all fireplaces aren't Rumfords," he offered with a shrug.

Next we walked a short distance from the house. "Let's look at something," I said, leading the way.

We stopped beneath a natural canopy of hemlock trees just to one side of the driveway. A thick layer of pine needles and a tarp covered a shadowy shape about the size of a phone booth that some fraternity pranksters might have tipped onto its side. A second, smaller boxy form rested a few feet away.

Numerous pools of brown water emptied as we lifted the tarp off to reveal almost thirty doors, standing on their edges, aligned like a deck of giant playing cards. They were all different sizes and colors, some with peeling paint and the remnants of broken hinges.

"Got 'em from a salvage dealer who was going out of business," I explained.

"For the house?" Ralph asked, only half needing an answer.

"Those are all antique window sash," I added, gesturing toward the second stack.

The wooden doors were resting on a makeshift base of scrap wood to keep them off the damp ground. We tipped the stack of doors to reveal a mantel at the rear, then lifted the mantel clear, setting it against one of the hemlocks so we could look at it.

"Could you fit a Rumford inside that opening?"

Ralph studied the mantel. He knelt down and looked some more. He measured its width and height with his hands.

"Yeah," he said finally. "I think we could make that work." His manner was deliberate. Perhaps it was at that moment that the job was his.

Yet my mind might have been made up a few seconds later when, like a trawling fisherman, I tossed out, "Maybe we ought to talk about a Russian fireplace, too?"

Ralph's expression seemed to somersault: from serious as a parson, he went to wide-eyed surprise and then to a broad smile. Then the words came out all in a rush.

"You know, I've never built a masonry heater, but I've always wanted to. I was gonna build one for myself but, well, renting and all, it's never quite worked. But I've done the plans. They're great. It'd be fun. . . ."

Ralph was definitely the man to build our chimney mass.

THE BUILDING BUSINESS HAS always had a love-hate relationship with technology. A carpenter we know once complained about a renovation he was doing on an early-nineteenth-century house. He'd been hired because of his restoration expertise, but given the owner's demands for creature comforts, fixing up the old and original became secondary to installing the new. The master bedroom complex alone had four phone lines as well as satellite television and multiple speaker cables. New partitions were required for his-and-hers dressing rooms, closets, and bathrooms. Both baths had marble floors, and one had an oversize whirlpool tub. A highly sophisticated lighting system required dimmers, spots, and low-voltage fixtures. Finally, there was a new fireplace in the bedroom (not a Rumford, by the way) where none had previously been.

"When I started in this business," my friend commented, "fixing up a whole house was less complicated. It didn't take seventeen subcontractors. Just me, a plumber, and a wiring guy. And in a pinch, I could do it all."

While Betsy and I weren't guilty of technological overload, we certainly could be accused of being overly self-conscious about where our house was to fit into the continuum of time. The chimney stack was a case in point.

First there was to be the Rumford. The arguments in its favor were that it was to function as a fireplace and as an emotional focus in our home. There's something primal in the lure of an open fire. Watching a fire burn is, when you think about it, about as dramatic as watching paint dry, yet somehow it's mesmerizing. There are lots of explanations for its atavistic appeal, and I for one am inclined to buy into the theory that in some corner of our collective memory there is a stock tableau: the fire burns at the mouth of a cave, while our ancestors are warm and safe inside and wild beasts prowl in the dark night outside.

Vitruvius took this line of thinking a step further. He credits the discovery of fire with producing nothing less than human society *and* language. As he imagined the events, lightning struck, and the human race, then no more than wild beasts, was at first terrified. But as the fire faded, the men approached and discovered the flames offered warmth. They added fuel to the fire to sustain its comforting presence, initiating humankind's first cooperative enterprise. Managing the fire and other evolving rituals also produced the need to name certain common objects and actions. "Therefore," concluded Vitruvius, "it was the discovering of fire that originally gave rise to the coming together of men, to the deliberative assembly, and to social intercourse." And eventually to shelters, like his primitive hut, that are the precursors of our house and yours. (Vitruvius himself might be interested to know that the word *focus* as we use it today evolved from the Latin word that he used to mean hearth.)

Call it sentiment, if you wish, but the fireplace with a crackling fire is comforting, familiar, and familial; the Rumford would give us an open fire to enjoy in a (relatively) efficient fashion that was also historically appropriate and environmentally sound. So ran our rationalization. Or part of it, at least.

A bird flying over our imagined chimney stack would peer down into not one but a line of three flues. The plan was for the Rumford to be served by one of them; another flue would carry the smoke away from a modern furnace; the third would vent a wood-fired masonry heater, the Russian fireplace that Ralph was so keen to build. To Ralph it presented a technical challenge, but I imagined that a masonry heater would be highly practical while serving a range of other philosophical needs.

One such need was to link our house with those of early British, Dutch, and French settlers in North America. The shock of the cold—they had rarely known freezing temperatures back home—had meant they had to rethink their building practices. One highly intelligent adaptation they made was to build houses with rooms that huddled around the chimney mass like scouts around a campfire. Most rooms in, for example, the durable Cape Cod house, had a fireplace, and each firebox was set into the masonry stack at the center. The brick mass Ralph would build at the core of our house would function in the same way: together the masonry heater and the Rumford (with the narrow flue for the oil burner passing between them) would be a giant radiator. I liked that historic association.

I had a parenting rationale as well. In my childhood home, I had a favorite heat register. In the winter, at age five or six, I would get up before anyone else in the house. Leaving the comfort of my bed, I would go directly to my spot in front of a louvered vent in the living room. Sitting in the rush of heated air was the perfect place to look at books, watch television, or simply observe the house coming to life around me.

That heat arrived as if by magic. As a young boy, I would not have been able to make any clear associations in my mind between the oil truck's arrival, the furnace cycling in the cellar, and the heat that made me feel so secure. In adolescence and beyond, in the sixties and seventies, I was astonished to learn that such comfort and convenience had a price (just as so many of my elders were stunned to discover that they couldn't take petroleum for granted as an unlimited and inexpensive resource). With the indignation unique to the teenage years, I was outraged to find that we squandered gas as we drove about in our enormous automobiles and burned large quantities of oil in overheating our uninsulated houses.

Perhaps the sting of realization from those days is still with me, since, in thinking about our new house, I wanted my children to understand that houses aren't automatically warm. To leave them ignorant of the environmental and economic complexities of energy just wouldn't do, yet I had no desire to act the role of pedagogue, like Ross Perot with his charts and pointer. On the other hand, I thought, if they could watch me carrying armload after armload of wood every day, feeding the stove to keep the house warm, maybe they would understand at a practical level something about our lives and our world. Who knows? Over time they might even assume some responsibility for the task themselves.

The logic of constructing a masonry heater wasn't based entirely on idealism. Betsy and I had spent the previous decade feeding the woodstove in our cottage. The relationship between heat calories and exercise calories had become more than a matter of half-forgotten laws learned in physics class. Experience had demonstrated the truth of the cliché that heating with wood heats you twice—once when you split the wood, and once when it burns. We had grown accustomed to the rituals of burning wood. Mostly it had been my job and an agreeable one.

Relying on a woodstove as a principal heat source did have a downside. The stove required more or less constant attention; someone had to monitor the fire and add fuel as necessary, typically at two-to-four-hour intervals. That meant a great deal of wood had to be cut, stacked, and carried. The flue had to be cleaned at least annually to eliminate the sticky, flammable creosote that would accumulate there.

Anyone sensitive to environmental issues must also feel a twinge of conscience now and again about burning wood. Although wood is a renewable resource, its combustion releases a telltale plume of brown smoke up the chimney. Today's woodstoves produce lots of unburned gases and particulate matter, together with an acrid odor.

The masonry heater was our secret weapon. Masonry heaters made of brick or whitewashed stone have been in use for centuries in Finland, Poland, and other northern European countries. In the alpine regions of Germany, Austria, and Switzerland, tall masonry heaters covered with tile also have a long history. But the Russian fireplace was the variation that would inspire our masonry heater. Its guttural Russian name, *grubka,* suited a large, squat pile of stones or bricks with a fire burning inside.

Whatever its name or country of origin, a masonry heater is the technological antithesis of the master bedroom complex my contractor friend described. In that suite of rooms, a radiant heating system in the bathroom floors involved a network of pipes, tubes, and valves that were, in turn, tied to a boiler in the cellar below. Ductwork and an air handler were required for converting more hot water to forced air. There were also air-conditioning components located in the utility room in the basement. A range of wires and cables, switches, thermostats, and other fittings were assembled into sophisticated control systems.

In contrast, our masonry mass, with its Rumford and grubka, would be built of bricks and mortar. All we would need to run either of them would be matches, yesterday's newspaper, tinder of fallen twigs and branches, and cordwood culled from the thick woods around us. Our masonry mass would be decidedly low tech: the control system would consist of how much wood we put in.

A masonry heater is, in a sense, a built-in woodstove. Both consist of a closed firebox with a door through which the stove is fed. The fuel burned is wood, and the heat from the fire warms the enclosure, which, in turn, radiates to the living spaces around it. The smoke is vented via a chimney flue. There, however, the resemblance between a grubka and a familiar woodstove ends, as a masonry heater diverges in both function and construction.

When in daily use, today's state-of-the-art *airtight* woodstove burns constantly. Its efficiency depends on regulating the flow of oxygen to the fire. Airtight construction and a system of vents allow the oxygen available to the fire to be reduced so that the fuel is consumed more slowly (and therefore burns for a longer time). By opening the vents, more oxygen is allowed to enter the firebox, and the output of the stove is increased. It's a kind of manual thermostat.

If a woodstove sounds unsophisticated (after all, it's nothing more than a fire in a box), then a masonry heater is in some ways an even more primitive device. The flow of oxygen to the fire in the grubka isn't controlled. The grubka fire burns very hot, reducing the fuel to nothing but ash within an hour or two. When the fire is out, the damper is closed, preventing any heat from escaping up the chimney. Yet the heating cycle continues.

A masonry heater in full operation contains no fire for perhaps 80 percent of the day. Sound like magic? Well, a cast-iron woodstove begins to cool as the fire within fades; after the fire is out, a metal stove is cold to the touch in an hour or two. In contrast, a masonry heater holds its heat for many hours after the fire burns down.

The circuitous path the smoke follows makes it possible for the grubka to hold heat for hours. The smoke doesn't simply ascend out the flue, as it does from a woodstove, but zigzags across, up, around, and even down within a maze of baffles, throats, and runs. The smoke gives up its heat before it reaches the chimney, warming the interior bricks, which in turn warm the entire masonry mass. After the fire goes out, the thousands of hot bricks continue to radiate heat. Think of it this way: in some cultures, a heated brick is wrapped in a cloth and put into a bed to warm the sheets. The grubka uses the same concept, only its several tons of brick mass retain vastly greater amounts of heat that can warm an entire house. It's really a heat-storage system that also happens to have a wood-burning firebox at its core.

When I first learned of masonry heaters, they seemed too good to be true. We were told the grubka would be environmentally friendly. Combustion takes place at such high temperatures that a masonry heater produces no creosote or soot buildup. The principal by-products would be carbon dioxide and water vapor, and there would be essentially no unburned gases and a minimum of particulate matter, meaning the heater would be almost smokeless. We would use roughly a third of the wood an equivalent woodstove would require. And the heater would have to be fed its diet of wood only twice a day instead of around the clock.

It made sense to me. On the other hand, Ralph couldn't predict how much heat a grubka would generate. He was confident it would be a lot—but enough to heat our twenty-five-hundred-square-foot house on a bitterly cold winter day? He didn't know.

For that reason and others, we planned on a third and final flue. We needed a reliable source of heat to supplement the masonry heater on the coldest days. Two young children, one of whom would barely be talking when we moved into the house, needed a warm, healthy environment. We required a system that could keep the pipes from freezing when we were away from home. It had to be economical to install and to run.

We considered the choices. The components of an electric heating system are inexpensive, but we ruled out electric heat because the annual energy costs would have been at least double that of any other choice. Municipal natural gas wasn't available in our rural area, so that simply wasn't an option. I liked the notion of a geothermal or heat pump system, one that would use the constant temperature of the soil or groundwater as energy for heating and cooling. But a geothermal system would have been very costly to install and, in our cold climate, not quite up to the heating task in the winter (heat pumps aren't very efficient when the thermometer drops below thirty degrees Fahrenheit, requiring the installation of a backup system, typically electric baseboard heaters).

Our research brought us to the conclusion that the best choice would be a hot-air furnace, together with a network of aluminum ducts that would bring the heated air to registers strategically located in the living spaces. Dollar for dollar, oil offered the most Btu. So the third flue would be a straight, tile-lined chimney that would vent an oil-fired, forced-air heating system. We hoped the grubka would provide most of the heat, but the oil burner was insurance that the house would always be warm.

I requested bids for the job from several HVAC contractors

who had done good work for friends and acquaintances. One of the advantages of living in a sparsely populated area is that word passes quickly when the work is good, and even faster when it's shoddy and unsatisfactory. These contractors helped me refine the plan. One zone for the house would be adequate, we decided, and central air would be a luxury in the Red Rock miniclimate, where we pined for air-conditioning on fewer than five nights in a typical summer. The estimates varied from just under seven to over eight thousand dollars.

When the costs of Ralph's work and that of the furnace were added, the total came to roughly $23,000. I realized that, at a stage very early in the process, we would have to pay out almost half the estimated construction cost of $85,000 for just three budget items: the masonry, the HVAC, and the foundation. But the elements all seemed somehow inevitable.

RALPH HAD VISITED THE work site when Mark and I were beginning to frame the first floor. We had agreed to financial terms by phone, so when he arrived, we both signed a simple agreement he had drafted. A check for $2,500 was due on signing. Most of that sum, he confided, would go almost immediately to suppliers. Bricks and concrete block would be arriving in a few days.

Ralph had also brought his tape measure and a couple of sticks of yellow chalk. He descended into the cellar.

In order to distribute the tremendous weight of the chimney mass—it would eventually weigh more than five tons—the plan called for a special concrete footing. This ten-inch-thick pad, lined with steel rebar, had been poured and subsequently hidden by the four-inch-thick concrete floor.

Ralph consulted the foundation plan to establish the location of the footing. We heard the familiar rattle of a tape measure being shifted from one spot to another.

"You need a hand?" I called down to him through the ladder hole.

Ralph looked up, squinting at the sunlight over my shoulder. "No, I'm almost done," he replied.

He left a few minutes later. After he had gone, Mark went to look at what Ralph had done. There was a large rectangle on the cellar floor. The powdery chalk marked where the chimney would rise. "I suppose its a start," Mark observed. He didn't sound very impressed.

Ralph's next appearance produced more visible progress. This time he arrived with his brother Dickie, a thinner version of Ralph topped with a baseball cap. They were in Dickie's truck, a dilapidated old Ford with a homemade wooden bed, which restored my faith in the contractor-pickup cliché. It held a masonry wet saw and several yards of mason's sand. The Bruno brothers muscled the saw off the truck and settled it under a giant oak tree.

They leveled the tool, positioning the blocks beneath the base. It consisted of a large, motor-driven circular saw mounted on a special frame. The blade was suspended over a water-filled trough. The saw housing was fixed, but a movable cutting table allowed bricks or stone to be presented to the blade. When in use, a jet of recirculating water kept the blade cool.

No sooner had Dickie cut one brick to test the saw (it worked) than he unplugged the motor. Ralph set to work and, in a few seconds, had removed the blade. He was wrapping it in an old velour shirt when he noticed Mark and me looking on quizically. He explained, "These blades are diamond tipped, expensive, and hard to get." As an afterthought he added, "Once I had one walk away from a job."

Ralph and Dickie shoveled the sand off the truck and covered it with a large mortar pan. Then they were gone again, the saw blade stowed carefully in the cab of the truck.

The next morning a delivery truck laden with masonry supplies arrived. Since that first pile of lumber Mark and I had restacked our first day, we had consumed several more enormous loads of materials. The same driver had made each delivery, and I learned his name was Dennis. Having unloaded the most recent supply of plywood and lumber, we talked about the weather and the relative pace of deliveries to nearby building sites. As I signed the receipt, Dennis said, "It'll go up quick from here." His words left me wondering whether he thought that we had been going slowly or that framing jobs always gained speed.

Another small remark he dropped almost as an afterthought had gone right over my head when he said it. "Maybe you'll want your next load brought on the boom truck," he had suggested. The boom truck reference only made sense when the masonry supplies arrived.

A boom truck has a cab at the front, a flatbed at the back, and an onboard crane in between. The crane is hydraulic, powered by the diesel engine of the truck, with a large arm or boom. The driver from the masonry supply yard brought five cubes of brick, each of which consisted of about five hundred bricks neatly stacked on top of a wooden pallet. At the end of the boom was a two-fingered fork like that on the front of a forklift. Using the control levers, the driver maneuvered the boom's fingers into the cavities beneath each pallet, lifted the load off the bed of the truck, then set it down next to the foundation. Two more pallets contained cement blocks. It was less work for him to off-load tons of bricks and blocks than it was to carry a week's groceries into the house. The boom truck, I vowed, would have more uses than one on our work site.

The next day Ralph and Dickie were already at work when Mark and I arrived at about seven-thirty. The mortar pan had a thick mix of gray mortar, and the first course of cement blocks had been laid. Ralph was steadily laying up blocks on the next course, kneeling on the floor. Dickie was tending, bringing Ralph mortar and block. The chalk marks had disappeared, and in their place a much more impressive masonry construct was just beginning to rise. It rose quickly—less than three days later, the cement-block structure reached the level of the first floor. Ralph and Dick then poured a reinforced concrete slab that capped the base. Beneath was essentially a hollow pedestal that would be an ash pit for the grubka and the Rumford. Above would be the two fireboxes, integrated into a single mass of brick.

With the base in place, the work on the fireplace and masonry heater could truly begin.

A FEW DAYS LATER, Mark was on his way to the cellar with a large cardboard box. The box was empty, but we planned to store wood scraps in it for later use in the fireplace and masonry heater.

Ralph stuck his head around a nearby partition. "Can I have that?" he inquired.

Mark shrugged and handed the box to Ralph, who tucked it out of sight in his stack of tools and materials.

The two of them together made a striking contrast. Ralph had the perfect build for a mason. He was of medium height, big boned, and muscular. Mark's long, lean physique looked quite out of place next to Ralph and his pile of brick. Masons are rarely tall and skinny.

The thought occurred to me that it's really a matter of

mechanics. Tall people have a higher center of gravity. They have to bend farther to reach the ground, and they must lift objects higher to reach their comfortable working zone. When they extend their longer arms away from their bodies, there's more stress on the muscles and bones.

As a tall man, I got a vicarious pleasure watching Ralph lay up brick with his compact frame and denser muscle mass. For him the work was all short strokes and economical movements. There's always pleasure in watching a good workman plying his trade, but here was the added satisfaction of seeing the red brick and the gray mortar. The concrete block he had laid up below had had little aesthetic appeal for me, but the warm red hues of the brick made the Rumford and the grubka seem more real.

Later that week, we found Ralph on all fours, atop the cardboard he had taken from Mark. It was nearly quitting time, and Dickie was cleaning mortar pans outside.

Ralph had opened the seams and spread the box flat on the plywood floor in front of where he was building the Russian fireplace. With a yardstick and a carpenter's pencil, he measured and marked a series of lines. Using a utility knife, he then neatly cut four large rectangles.

He had become accustomed to finding Mark or me standing nearby wearing a questioning expression. He pointed into the heart of the brick structure that was becoming the Russian fireplace and began to explain what he was doing.

Every conversation about masonry heaters seems to begin with the fire. After all, the intensity of a grubka's fire is what distinguishes it from a Rumford, a woodstove, or any other wood-fired heater. The masonry heater is designed to produce heat rapidly by burning at very high temperatures, typically in the range of two thousand degrees Fahrenheit.

The fire needs to be contained within the firebox. Since common red brick will crack, spall, or even melt at such high temperatures, Ralph explained, he was using firebrick in constructing the firebox. The clay used to make firebrick is a grade referred to as fireclay, which will withstand temperatures in excess of three thousand degrees Fahrenheit.

The firebrick would expand as it got hot. That was where the cardboard came in. Ralph had just completed the firebox and was about to build the redbrick shell that, like the containment building at a nuclear power plant, would enclose the firebox. There needed to be enough space that the hot firebrick, swollen with heat, wouldn't crack its masonry shell. Yet the covering also needed to hug the firebox closely so that the heat would be transferred to the exterior of the masonry mass.

Ralph positioned the cardboard around the firebox. Once it was in place, he explained, he would build the redbrick shell. Later, after the whole superstructure was finished and the chimney capped off, the Russian fireplace could be fired. The hot brick would reduce the cardboard to ash, leaving a thin, uniform space—an expansion joint—that would allow for the expansion of the firebrick. There would be no risk to the brick box around it, yet the heat could pass directly from the firebox to the surrounding shell.

Imaginative solutions always impress me, so I nodded appreciatively as Ralph told us what he was doing.

Mark was still standing nearby. "Have you ever done this before?" he asked.

"No," said Ralph quickly. "But I'm pretty confident it'll work."

We were all silent for a moment. I considered the mix of brick and cardboard before us. Ralph busied himself applying the last strips of tape to the cardboard now in position around the firebox.

As I started toward the ladder to go back to work on the floor above, Mark followed.

"Some things you've got to take on trust, Mark." I said it because I believed it: I had hired a good man to do his job, and now he was doing it.

Mark nodded, but he didn't look convinced.

ONE HOT AFTERNOON, THE masons left a little early, but Mark and I worked on. Our conditioning and stamina were improving, and a few previous days with cooler temperatures—in the seventies—had added to our energy reserves. We were both feeling confident even as a hint of autumn was in the air.

The house was rising steadily. The second-floor exterior walls had been completed. We were framing the interior partitions as we readied to hang the ceiling joists. The rafters and roof would follow.

All in all, we were running close to schedule and we had a feel for the rate at which we were working. In another week, the framing would be two-thirds completed. Although Mark would be heading back to England only a week after that, we had a shot at completing the box before he left. The roof wouldn't be on, but we wouldn't be far short of it. Some temporary help could be found to finish closing up before winter.

That humid August afternoon we were about to nail a partition wall in place. Mark had been pounding the sill plate into position using the persuader, driving the sledgehammer's head at the bottom while I balanced the top. The hammering jarred a piece of two-by-four off the top of the partition and into free fall. Accelerating rapidly, the two-by-four fell through the hole in the platform where the chimney would soon rise. It was then lost from sight, but a moment later, we heard a thud as the piece crash-landed. By sheer good luck, no one had been in its path as it fell.

I bent over the hole to look down. The masonry mass had taken a direct hit, and the chunk of wood was cradled by the sheet of canvas Ralph had used to blanket his day's work.

I immediately imagined hours of labor on the masonry heater ruined. We climbed down the ladder to the first floor.

With its covering still on, the masonry heater had the shape of an igloo and didn't look like two solid weeks of work had been invested in it. The piece of wood that had fallen had created its own imprint as it sat awkwardly. We removed the cover and looked for the point of impact.

"Not that I know what I'm looking at," said Mark, "but it looks okay to me."

I nodded tentatively.

"But what was that clanking sound after it hit?"

Mark saw the level first. It was an eight-footer, black, and made of aluminum. Ralph used it as a straightedge to keep his work straight and plumb. Mark pointed to it, his eyes wide. The level, although it had been made in the rigid shape of an I beam, had bent as though it were no stronger than the tine of a fork. The end of it bent away from its original bearing like a sharp turn in a road. We both flashed on how much force that chunk of scrap must have had, but then Mark started to laugh, and I found myself laughing, too. It was as if someone had surprised us with a pratfall.

Our laughter was a release. We certainly knew we were doing dangerous work and tried to anticipate the risks. But we both were smart enough to know you can't think of everything. This was the closest we had come to a real accident, and the only harm was a ruined twenty-dollar level. It was some sort of friendly warning we could laugh about.

Having Ralph and Dickie on the work site was a sociable change. We would eat lunch together most days, and stories would get swapped, work notes compared.

We learned that their paternal grandfather had emigrated from Italy shortly after World War I. In his early twenties, he came to America to make his fortune. He caught on with a construction company, learned the mason's trade, married a local woman, and settled down in western Massachusetts. Ralph and Dickie represented the third generation of Brunos in the trade, as their father, too, was a mason.

Their family history worked to our advantage. After finishing his sandwich one lunchtime, Ralph went to his car and opened the trunk. He pulled out a block of white stone and brought it back for the rest of us to examine.

The stone was about a foot square and perhaps five inches thick. It was white marble and had been tooled around the edges. It looked like it had been salvaged from an old building.

"An antique sample, Ralph?"

"No," he said, and laughed appreciatively.

I didn't know I'd made a joke.

"I did it last night."

I looked at the stone again. On closer examination, the tool marks did look as if they might have been quite recent, but the block looked well aged. Ralph intended it to be a plinth block for the mantel. Each side of the antique wooden mantel would sit on such a block, like columns on their bases.

"I've got a hearthstone, too," he added, "but it was too big to fit in the car. It's a Carrara marble."

I asked him where he got such stones, half-worried that the price would be going up. The hearthstone, which was one inch thick and two feet by five feet, had been a panel in the wainscoting of a grand Beaux Arts train station in Pittsfield. During the benighted days of urban renewal in the 1960s, the building had been demolished. Ralph, then a teenager, had been drafted to help his father salvage pieces of beautifully veined marble and cart them home to their backyard. There they had gradually been swallowed up by the weeds. The plinths and the stones that Ralph would shape into a flattened keystone arch over the Rumford's opening had been saved, too, from another demolition. They had been window lintels at a school.

Despite his family history, Ralph had come to masonry by a rather indirect route. He was nearing forty, so I was surprised to learn that he had been working as a mason for only eight years. That left a decade unaccounted for, but the conversation had moved on and the question remained unasked.

A few days later, Ralph had run to his car just after five o'clock and virtually raced out of the driveway. Dick remained behind, hosing out the mortar pan.

"He was in a hurry, wasn't he?" I remarked.

"He's got a haircutting appointment," said Dickie as he shut off the hose. "He still keeps a few of his old customers," he added blandly, heading to his truck.

Betsy was the one who asked Ralph directly about his past career. He and Dick had just finished the brickwork on the first floor, which meant the Rumford and the masonry heater were done, leaving only the chimney to be completed. Betsy and the girls arrived with a bottle of champagne to celebrate. Sarah settled in to play with the stack of wood scraps Mark had collected for her as makeshift building blocks, while Elizabeth explored the straps that contained her in her baby carrier. Both were fascinated by the pop of the champagne cork and its flight out a rough opening in the wall.

We toasted, and pretty soon I heard Ralph telling Betsy he had been a hairdresser.

"Not many people make career moves from the beauty salon to the building site," Betsy said. He liked cutting hair well enough, Ralph told her, but he had always wanted to be a mason. His father and grandfather had discouraged him: "It's hard work," they had told him. "Your body wears out young." But he decided to find out for himself.

I got talking to Dickie that evening, and he was a surprise, too. He had arrived that morning looking very haggard. The thought occurred to me that perhaps he had been out having a few drinks and, with the arrival of the morning, had found himself with a hangover. He was smoking a Marlboro as we talked after work that afternoon, and I made some vague remark like, "How was your evening last night, Dick?"

Before long, he had told me of moonlighting as a guitar player in a rock band. The evening before, he had played at a wedding celebration that had lasted well into the night.

Betsy led Ralph to tell her more about himself. He talked about embarking on a second career after the age of thirty and told her that unlike his father or grandfather, who had served long apprenticeships, he had been by necessity in more of a hurry. He had taken a job with a small-town mason. We were all listening.

The man was roughly thirty years older than Ralph and was the masonry equivalent of a general practitioner. He had done fancy work for rich people and put block chimneys onto mobile homes. He'd laid patios, brick pavers, and tile. Most of the fireplaces built nearby over the last twenty-five years had his initials scribed somewhere high in the firebox.

Ralph learned all he could, but the man was no teacher. He never stopped to explain anything to Ralph because he had done it all himself so many times before that he didn't think much about it. He knew every shortcut there was, too. Not surprisingly, his work was slapdash, a trait that made Ralph very uncomfortable, but Ralph got his apprenticeship in a few short months before moving on.

Some of what he learned had to do with customer relations. He told us a story about one client who carped about everything. "The guy didn't like the color of the brick, even though it was his choice in the first place," said Ralph. "He complained the fireplace hearth went too far into the room, though that was just what the building code required. The job was going too slowly, he said, although that wasn't our fault—we'd been delayed by the carpenters. Every time the client came to check on progress, he had another complaint." Ralph shook his head, mildly irritated even ten years later.

His boss, Ralph reported, began to experience a sense of financial foreboding about this client. He didn't need a crystal ball to sense that his final payment might be withheld unreasonably. One morning the client arrived, griped about how far over budget the house was, and left again. Then, Ralph noticed, the mason went to his truck and returned with a flat parcel wrapped in brown paper. He unwrapped it, revealing a pane of glass.

"I was curious," Ralph said, "but I didn't say anything. We'd never used glass before, but I understood there were lots of tricks I didn't know." Ralph smiled slightly, then continued with his story.

"My boss climbed to where we were laying up the chimney. He set the glass neatly across the flue. Then we just went back to work, being careful that no bricks or globs of mortar fell into the flue. But he never said a word about the glass. We finished up and moved on to the next job."

The bill, Ralph told us, was then duly submitted, and as the mason had anticipated, the client complained bitterly. Ralph learned this after the fact, when he and the mason made a detour a couple of weeks later to the cranky client's house. As always, the conversation with the client began with a complaint, this time that the chimney didn't work at all.

"Yeah, well, and it won't until you pay my bill," the mason assured him gruffly.

When the client began to yell, the mason—who was a physically imposing man—took one step toward him and gestured with his powerful right hand, palm out, signaling *Stop!* The intimidated client suddenly went silent.

"You pay my bill, and your fireplace will work. You don't, and it won't. Period."

Almost immediately, the client started to complain again. The mason turned and walked away. Ralph followed him. They heard the client's voice trail off. He called to them as they were reaching the truck.

"All right." It was a whine. "Let me get my checkbook."

A few minutes later, the check was written. The mason pocketed it. He instructed Ralph to get a ladder from the truck and lean it against the house. Then the mason climbed to the peak of the roof with a half-brick in his hand.

"He was all business," Ralph explained. "This wasn't a ceremony. He just went over to the chimney, looked down to make sure it was the right flue, and dropped the brick."

Even outside the house the sound of shattering glass had been audible.

By the time Ralph delivered it, Betsy and I knew what the punch line was going to be, but we laughed appreciatively.

Ralph picked up the thread. He and the mason had loaded up the ladder and left without a word. His boss's expression, which had remained a stoic mask throughout the visit to the house, began to change only after they drove away. Ralph chuckled appreciatively as he remembered.

"He had the biggest smile on his face I ever saw."

LABOR DAY HAD COME and gone—we took off only an afternoon, since Mark's time was dwindling. We were still many weeks from winter, by which time we simply *had* to seal up the box to keep the weather out. Though we were closing in on completion of the second story of the main block, that still left the rafters on the wings and the attic roof, and then the roof itself. At some moments, it seemed impossible. At others, I felt confident we could do it.

At about six o'clock one evening, only a dozen floor joists remained to be hung to complete a major section of the attic floor platform. Mark and I were taking stock, standing in what would eventually be the upstairs hall.

"Shall we finish this line of joists?" I looked upward at the sky visible through the unframed opening overhead.

Mark put down his water bottle in one of the bays between the studs. He was already back to work, reaching for one of the joists. I moved to grab the other end.

"Won't take long," he said, two steps up his ladder.

The next few joists went up smoothly. Then we made a mistake, nailing one on the wrong side of a pencil line, meaning it wasn't parallel to the others. We had to stop to correct the problem, but a few minutes and many curses later, we had loosened and reattached the joist in its proper place.

We were working our way down the length of the hall, hanging three joists, then moving our ladders to reach the next several. At one position, we tried to hang four before resetting our ladders, meaning that Mark had to reach well to his right to hang the last of the set.

His stepladder wasn't opened like a letter A but was angled against a stud wall. Mark was perhaps four feet up, his arm extended straight out as he hefted his end of the ten-foot-long two-by-ten. As his weight shifted, the ladder started to move.

The ladder slid from beneath him as if in slow motion. He

tried desperately to keep his balance, but gravity pulled him down. The heavy joist was falling, too, its line of descent directly at Mark's head. I was overextended, almost helpless at the other end of the joist, which twisted out of my grasp. Mark tried to jump clear and ducked beneath the falling joist. It missed his head and seemed to slide over him. With a crash they both hit the floor. Then it was suddenly quiet.

Mark had landed on his feet, knees bent, arms raised to protect his head. He stood slowly.

"Are you all right?" I called.

"I think so," he said. He rotated his arm, rubbing his shoulder, and grimaced. "It glanced off here."

I quickly climbed down my ladder.

"I'm okay," he said. "But I don't know about Mr. Nail Gun."

I followed his gaze. The joist had scored a direct hit on the compressor. If it was broken, Mark had realized instantly, there would be no nail gun until it was fixed.

We bent to look. A carrying handle of rugged pipe had been cleverly positioned to shield the delicate copper piping and the rest of the works. But the joist had bent the handle, which had been jammed onto the pressure gauge. The gauge, too, looked bent, but there was no audible hiss of escaping air. In a minute or two of checking and straightening, we decided the compressor had escaped serious damage. "We'll check it again tomorrow," I concluded, "but I think we got lucky."

We took off our tool belts, covered up the equipment with tarps, and walked off the work site. I shook my head as we headed to the cottage.

"I'm sorry, Mark."

He nodded his head but said nothing.

"That was stupid," I went on. "We should have quit when we misnailed that joist. It was time."

He didn't say anything.

"People get hurt when they're tired," I added lamely.

We walked in silence, but as we approached the house, Sarah came running out to greet us. "Markie!" she called out. "Daddy!"

Mark had a relaxed smile on his face as he took the hand she offered him. She reached up to me with the other, and as her little hand disappeared in mine, a powerful recollection flashed to mind.

THE MONTH WAS NOVEMBER, almost three years earlier, and Betsy was pregnant. Her first contraction struck at about 10 P.M. on a Friday night.

We had been eating a leisurely dinner in a restaurant, half of a foursome with another couple. Betsy wasn't sure she was in labor, so she kept her symptoms to herself. An hour or so later, she broke the news as the two of us drove home.

"I think the baby's coming," she said, as if she were offering an offhand remark about the evening we had just passed. Nervousness can produce such lightness in Betsy at moments of tension.

I was nearing forty, and she was three years younger. Together more than ten years, we had pursued our careers, unsure whether we would have children. Once we decided to, it had taken more than a year for Mother Nature to get us on course for parenthood. During the nine months that followed—nine months to the day, as Betsy's labor began on her due date—we felt we were readying for a trip, and the sense of expectation agreed with us.

By 3 A.M. Saturday morning, we were walking down the hospital corridor on our way to the labor and delivery rooms.

In the course of that day, our world changed. For the better, I might add. As parents-to-be, we had known we were moving on to the next stage of our lives. Betsy had had a healthy pregnancy, having felt surprisingly well throughout. We had performed the requisite nesting tasks, accumulating the necessary clothes and equipment.

We were both in a state of heightened anticipation: expectant parents are no longer childless, yet not quite parents. It's like the greenroom in a TV studio, where the guests wait to be summoned to make their appearances. You've made it; you're almost there. You're about to face the camera and experience your moment of fame. You've been told what to expect, yet there's a strangeness about the impending event.

Throughout the night, Betsy's obstetrician came and went. We saw the gray light of dawn arrive after 6 A.M. The labor was proceeding slowly, but a delivery nurse named Jane helped us bide our time. The three of us did the crossword puzzle from Friday's *New York Times,* and then Saturday's puzzle, too, from a paper I bought at the hospital gift shop.

Betsy asked the doctor on a midmorning visit what would happen if the pace of her labor didn't quicken. He replied that it might be about time to think about induction—that is, to help things along by injecting Betsy with a pituitary hormone to stimulate the uterine contractions. Betsy had heard other new mothers recount the discomforts of induced labor, and the power of suggestion seemed to have a magical and instantaneous effect. Her contractions immediately increased in intensity and frequency.

Not long afterward, the time came for Betsy to begin pushing the baby out. A swarm of nurses and orderlies moved the mother-to-be into the delivery room, with me trailing close behind in appropriate hospital garb. The delivery nurse, by then a full partner in the process, stood at Betsy's right while I held Betsy's left hand. Jane and I coached her, counting to guide her breathing and her pushing.

Three doctors stood by, including two residents and Betsy's obstetrician, a young man who bore a striking resemblance to television's onetime boy-doctor, Doogie Howser. (That's one of the disorienting parts of being fortyish—all of a sudden your doctors are younger than you are.) I watched the physicians between contractions. They were perhaps ten feet away, positioned so as to be able to see Betsy's pelvis and watch for the baby's head to appear. They talked quietly among themselves like three tourists at a bus stop. They were solicitous but detached.

I wasn't detached, believe me. The events were wonderful but scary, although the presence of nearly a dozen medical professionals going about their respective jobs with such utter calm was reassuring. Betsy was experiencing tremendous stresses, which I could do little to lessen. I felt like a passenger being taken on a ride unlike any he had ever been on. Who was driving the bus wasn't quite clear, although in Betsy's state of pained exhaustion, she didn't seem to be.

About twenty minutes into the pushing, Betsy and the baby again seemed to have reached a stalemate. The baby had traveled partway down the birth canal but wasn't progressing. The obstetrician mumbled something about forceps. Perhaps that's an accepted strategy. I can imagine one doc saying to another in the scrub room after a delivery, *I used the old get-out-the-forceps trick on her . . .* , and the second doc responding with a knowing nod and an appreciative chuckle at the other's guile. The power of suggestion worked again. The mention of forceps got things going, and within moments Sarah emerged.

I was unable to speak after the birth. My throat was swollen with emotion. If I had been brought up to reveal more of my feelings, I might have cried freely. However, keeping emotions within is the way of my world. The power of the moment wasn't diminished one milliampere by the fact that the mini-implosion took place inside my head and heart. And body, by the way. The strain I felt wasn't in any way comparable to Betsy's. Different thing. But the coaching, the watching, the counting, the rest of it . . . it's more than sideline stuff. It left me limp, a little shaky, and exhilarated.

Baby Sarah—pink, her head slightly and temporarily misshapen from her fight down the canal, her body still slick with the residue of the vernix, her umbilical cord clamped off—took to her mother's breast instinctively and nursed briefly. I held her for a moment, too, a nervous and novice baby holder. To cap off an afternoon thick with emotional surprises, I found myself almost overwhelmed by the desire to protect this child and her mother.

That was a feeling unlike any other I had ever experienced. It's one that has grown wonderfully familiar as, in some measure, I have reexperienced that sensation every day since, watching Sarah learn to climb stairs or ride a bike or just sleep peacefully in her bed.

I began to understand on the day Sarah was born that children are the engines that drive families. It had been Sarah, I know now, who had driven that bus. She also started us on the trip out of our small home. Elizabeth's arrival two years later had added to the momentum of things.

We had gone on to begin the work of building a new house. Our journey had taken us to our new property. We had welcomed Mark Lynas into our little world, and it was Mark, someone else's child, who had almost gotten badly hurt building our house.

That, I vowed, would not happen again.

The Negative Moment

moment (MO-ment), *n.*
Mech. : Tendency to produce motion,
esp. about a point or axis; *negative moment* : motion,
as in a structure, in the direction opposite to that
which is considered positive.

I have no fondness for high places. Not that I'm acrophobic (how could a devotee of classical buildings be afraid of something even remotely connected to that great city on high, the Acropolis?). If I set my jaw, put one foot on the first rung of the ladder, and climb without pause, no paralyzing fear overcomes me. Launching a house-building project like ours simply wouldn't have been possible if it did.

Working twenty feet or more off the ground does have occasional compensations. From the top of a tall platform—such as a half-built house—even a familiar scene becomes a new vista. The sight lines are wide and unobstructed by a safety railing.

The sky seems more expansive. But it isn't the bird's-eye view that I like best. It's the birds themselves. The realization that I was sharing their realm first came to me one day years ago as I was putting shingles on the roof of our previous house.

As with any unfamiliar job, there was the learning how to do it. A reference book advised using aluminum drip edge to trim off the perimeter. Next came the snow-and-ice shield, a bituminous membrane that ran across the lowest portion of the roof. Tar paper was then stapled over the rest of the plywood subroof. On top of this underlayment, the shingles were nailed down, their joints staggered from one course to the next.

Some planning was required to lay out the lines of shingles to make sure they were even and parallel. But the most difficult part of the process was mastering the coordination of working on a pitched roof. I tried kneeling, but the granules of the asphalt shingles scraped the skin of my knees through my jeans. The butt-down scooching motion of a crab walk proved inefficient. The gorilla walk emerged as the best option: facing up the slope of the roof, with shoulders hunched and knees flexed, you shift your center of gravity forward, achieving better balance. On a roof, stability is the first priority.

My back and neck began to complain barely an hour into the process, so I grabbed my water bottle and sat astride the ridge of the roof. Relaxed, waiting for the muscles in my neck to loosen, I realized I was being watched.

Not more than twenty feet away, in a branch of a tall hemlock at my eye level, was a medium-sized bird, bigger than a robin, with black wings, a white abdomen, and a rose-red breast so brilliant it seemed to shimmer in the sunlight. After a moment, the rose-breasted grosbeak made a short, almost mechanical *clink,* like the sound of one empty soda can striking another, and flew off.

Roof work still isn't fun. Heights remain intimidating, and there's less satisfaction in physical work when no more finesse is required than to align shingles and nail them in place. But my roof-time recollections do include the moment a blue jay flew by at a great rate, screaming its raucous screech—probably because it was being closely pursued by a Cooper's hawk more than twice its size. Turkey buzzards always seem to be floating high overhead, soaring in the thermal drafts. One day an even larger bird came tantalizingly close, not near enough for it to be a confirmed sighting of a bald eagle, but I'm sure that's what it was.

Up on the roof, one can experience the illusion of being in an aerie, getting brief glimpses into the lives of birds as the seasons change and the birds come and go. Each autumn we hear geese honking overhead, making their long journey from Canada to the Gulf Coast. The year that we began building our house, we heard the first gaggle flying south just after Mark himself had flown back to England.

AFTER MARK LEFT, THE autumn seemed anticlimactic. Not that work stopped, but there were fewer dramatic changes. The walls were standing, and the wings had a roof covered by a layer of plywood. But the rafters on the main building were not in place. They still had to be cut and nailed, along with the plywood subroof. That would give the house its final mass. Then the roof surface had to be shingled to keep out the snow and ensure a weather-tight work space for the electrical, plumbing, insulation, and other interior jobs to be done during the winter months.

Mark's exit left me with a personnel problem. No replacement was ready to step into his work boots; nor had Mark and

I reached any understanding about next summer. I had no more thoughts of a year hence than a fullback does of green pastures as he confronts giant defensive linemen intent upon throwing him bodily to the turf.

Other members of my family seemed to have more perspective on Mark's departure. Sarah wasn't yet three, but she asked at regular intervals, "When is Markie coming back?" Mark had gotten into the habit of scooping up eight-month-old Elizabeth after work each evening, calling her *Bébé* and cradling her in his arms and lap for hours on end ("I never knew I liked babies," he had confided, both warmed and intrigued by the discovery). When Sarah spoke the name Markie in the weeks immediately after he left, an expression would flicker across Elizabeth's face that was recognizably happy but also confused. As the mentions grew fewer, Elizabeth ceased responding to them. On the other hand, Betsy consoled herself with the secret knowledge that he would be back (secret because even Mark didn't know it was true). Unfailingly, she told Sarah, "Next summer. He'll be back next summer."

There was no obvious solution to my immediate need for help, but Betsy suggested I call a friend in the landscaping business. She knew that he would be experiencing his usual downturn as the cold weather approached and that he had two good men on the payroll. He would eventually have to lay them off but wanted to keep them employed through as much of the winter as possible so they would be inclined to come back to work for him in the spring. I objected at first ("What do they know about carpentry? I'm not running a trade school here, you know"). But her idea was a good one, since semiskilled labor could help finish closing up. I made the call, and a deal was quickly struck.

In a matter of days, my new helpers, Mike and Chris, and I

completed the last of the frame and nailed the plywood subroof in place. That meant the shingling could begin. Ralph and Dickie were still at the work site, and the chimney was just emerging from the peak. They were less than a week away from capping it off, and they constructed a homemade wooden scaffold to support them and their buckets of mortar and stacks of bricks. Mike and Chris began laying the courses of shingles at the eaves, working their way up the roof six inches at a time. Steady progress continued as October wore on and the weather grew cooler.

Each day I bought both crews lunch. Orders were taken at about eleven o'clock each morning, the list was phoned in, and then either Betsy or I would pick up the sandwiches. When lunch arrived, we all found seats—upside-down buckets, the end of a lumber pile, a piece of floor next to a wall. For the roofers and the masons, it was a time to relax, to rest the muscles and bones. For me, it was also an opportunity to get better acquainted with my new co-workers and, in particular, with Mike Beecher, whom I knew the least about.

I noticed that Mike ordered the same thing every day, a roast beef sandwich. He specified that it was to be prepared on a hard roll with salt, pepper, and mayonnaise. He would settle himself comfortably, open the butcher-paper wrapping around the sandwich, and eat at the same deliberate pace at which he worked. He liked food and knew a good deal about it from his days working as a short-order cook. He had had a lot a jobs for a man of twenty-six, having also worked cutting cordwood, as a stonemason, in an auto body shop, and as an apprentice electrician.

He ate with care and attention, drinking a can of Pepsi at the same time. He said little, speaking when spoken to, though listening carefully and laughing at all the right times with evident pleasure. He would finish by eating the pickle that came with the sandwich, then rise, saying "Thanks for lunch" with ritual politeness. Without another word, he would go back to work, even though fewer than fifteen minutes had elapsed since the sandwiches had arrived.

He didn't expect everybody else to do the same, but he liked being occupied, keeping his hands busy. Mike seemed happiest when he was going about his work in the same measured, deliberate manner he ate a roast beef sandwich. He didn't have Mark's impatience, but he, too, was happier when he was in action. We would watch him pick up his tool belt, buckle it in place, then walk away. We'd hear his footsteps disappearing down the hall, then the soles of his boots as they scraped the rungs of the ladder to the second floor. The sounds of work up above would soon resume. Work, steady work, was a fixing point for Mike.

One lunchtime the talk turned to the hunting season, which would begin in mid-November. Mike suddenly became the focus of the conversation. Not that he said very much; he still spoke little. But the conversation seemed to circle around him. He was an expert hunter and had the unspoken respect of the other men. They hunted, too, and successfully. But none as well as Mike did.

He talked that lunchtime about going hunting the previous season with the thirteen-year-old son of his employer. The boy seemed very eager to learn to hunt. Mike, whose native instincts had made such lessons unnecessary, dismissed the boy's efforts in a kindly voice. "I sat him down on a stump and found myself one. Then I watched him. He was swinging his gun around. He couldn't sit still for two minutes. Every deer within a mile knew he was there." He shook his head, bewildered by the boy's behavior. Then, as usual, Mike went back to work.

The picture I was left with was of a man who chose not to

kick back and talk to people after lunch, one who preferred to work to pass the day. But in the woods? There he found a quiet within himself that he knew in no other part of his life.

He was a good worker and an unassuming companion. But as October gave way to November, I had to leave Mike and the rest to their work. I was back to making my living as a writer. But I left with the confident feeling that I knew who I wanted Mark's successor to be.

SOME MONTHS BEFORE, MY phone had rung. An old colleague was calling to offer me a ghostwriting assignment. Ironically, the task that would take me from the work site was the writing of a book about builders' tools.

Betsy and I had discussed whether I should take the job. The income would be welcome, as the house was beginning to run slightly over budget. Increases in lumber prices meant that the framing had cost a few thousand dollars more than my estimates. Since others were doing the roofing work for me, that, too, represented an added cost, as did a worker's compensation policy covering Mark and me. A quick and dirty estimate put us almost ten thousand dollars over budget.

I reasoned, as well, that interrupting construction at this stage made sense: the added income would give me the freedom to work at the house nonstop later on, perhaps even carrying us to completion. Betsy and I agreed that it was a good plan, although taking a seat in front of my computer meant that my work at the house would cease and I would be watching winter approach from my office window.

The weeks passed quickly enough. We saw snow flurries early in November. Like thunder in the distance, those few light flakes fluttering down were a warning of things to come. One day shortly before Thanksgiving, clouds began drifting in at noontime. When the light that filtered through took on a yellowish gray cast, I felt the little annual thrill that still strikes me as the first storm of winter approaches.

The weatherman had been warning us for several days that snow was on its way, so I decided to celebrate and leave off working at my word processor for the afternoon and spend the hours at the house. It was partly by choice and partly by necessity.

Mike and Chris had finished the roofing, and a tight, triple-thick layer of asphalt shingles covered it from eave to eave. I'd asked Mike to work another day with me in order to help wrap the entire house like a giant birthday present, using a fabriclike covering called house wrap that came in nine-foot-wide rolls. We had worked together well, stapling the high-tech sheets of olefin fibers over the exterior walls. The wrap would serve the dual purpose of stopping wind from moving freely into the wall and allowing the escape of moisture, which, in the finished house, could accumulate in the wall cavities.

The house wrap initially covered the windows and doors, too, but I had gone back and cut and trimmed the wrap, stapling the loose flaps to the two-by framing inside. The natural light that poured through allowed me to see what I was doing indoors. But with a storm expected, those openings also represented giant holes that needed to be covered to prevent wind-driven snow from entering.

The solution was four-mil polyethylene, clear plastic sheeting the weight of sturdy garbage bags. I cut rectangles roughly a foot wider and taller than the window openings. The top of each sheet was stapled to a scrap of pine cut to the width of the window. The wood, which would act as a batten, was nailed to the top jamb of the window. The poly then hung in the opening like an oversize roll-up blind.

Working from outside, I stapled the bottom end of each plastic sheet to the wall below, then worked up the side of each window, stapling the poly to the sides of the opening. Fastened firmly, the poly would shed rain and snow and also let in lots of light.

The last of the window openings was closed just before the first snowflakes began to fall. I hurried to seal the doors with the blue tarpaulins we had been using to cover piles of lumber outdoors, since the supplies were now being stored safe and dry inside. The tarps were battened, too, across the tops.

By the time the openings were all covered, the snow was falling in earnest. The temperature hovered around thirty degrees, so the snow was damp and heavy, and the large flakes accumulated quickly. The scattering of leaves outside gradually lost definition, then disappeared altogether. As the snow continued to fall, the day darkened well before sunset.

I swept up the interior, organizing materials and tools. I peered out occasionally, gauging the rate of snowfall. But my mind was occupied: this was, after all, a moment of real accomplishment. We had managed to transform idealized paper drawings into a giant wooden sculpture. Indoors, the impression was of being inside a skeleton. Looking at it offered little sense of our grander aspirations; it was a sliver-and-shim practicum of the builder's trade, with its clutter of debris and equipment. But it offered shelter from the elements. For the first time, the great, hulking structure around me felt like a real building.

My breath was visible in the cold. The house was unheated, since the HVAC crew wasn't due to arrive for a few weeks to install the ducts and furnace. The Russian fireplace was finished, and Ralph had capped the chimney. (His final invoice had been for precisely the sum agreed upon months earlier, plus an extra charge of about thirty-five dollars for a replacement level; I gladly paid it all.) But he had warned me it would not be safe to light a fire in the grubka until the masonry had had more time to cure. The Rumford, though, was another matter. "You could start with a couple of small fires almost anytime," he had said.

I lit a crumbled sheet of newspaper to test the flue. Instantly the smoke was drawn upward. Ralph had left no glass in our chimney, and Rumford's design definitely worked. Fueled by wood scraps, sawdust, and a couple of brown paper bags left over from take-out sandwiches, a modest fire was soon set in the firebox.

The fir scraps crackled as they burned, but a great quiet had enveloped the job site. There were no screaming saws. The compressor was off. There wasn't even any conversation, since I was alone in the house. Outside, there was the muffled silence of the snowy woods.

JUST BEFORE FIVE O'CLOCK, I heard Betsy's car. She was on her way to pick up Sarah at day care.

Although a shy child, Sarah on more than one occasion had walked up to another child her age in a public place such as a grocery or department store. Without so much as a hi, she would embrace the fellow toddler, having never seen the youngster before, all the while smiling happily and talking nonstop.

"She needs other children," Betsy had concluded. She enrolled Sarah in a nearby day care run by a delightful woman who, looking much younger than her years, told us she had opened Long Barn Daycare because she was tired of waiting for her thirty-something son to produce grandchildren. Sarah was with Miss Diane that snowy afternoon as the first fire in the Rumford blazed.

I had planned to work into the early evening, but I suddenly had a desire to join Betsy in picking up Sarah. I got out to the car just as Betsy was climbing out. Elizabeth was asleep in her travel bed in the backseat. "I'll drive," I said.

Betsy agreed immediately. "It's pretty slippery, I think."

The thickening blanket of snow—by then it measured about three inches—didn't intimidate me. I learned to drive in a New England winter, and even as an adult I continued to find a certain macho pleasure in pushing the limits of a car's traction. There's a special thrill in taking a turn in a skid, almost but not quite out of control. For me, a normally cautious person, the first snowfall of the year provided a chance to test my anticipation, good sense, and ability to play the odds. On a back road, especially after dark, when the headlights offer ample warning of an oncoming car, the first snow drive of the year had become an annual rite of passage. Yet that day, with a baby in the car, I knew I would take no real risks.

We set off to pick up Sarah. Once out of the driveway, I tapped the brake pedal deliberately to test the slickness of the road. Conditions were slippery, but there was the satisfying crunch of the tires biting and the feel of the car slowing. We headed down the hill confidently but at a measured pace.

About halfway down, the road bears gradually right. Suddenly, our headlights were reflected back at us. A car seemed to be coming toward us without its lights on. Then the dark silhouette registered as a parked car. The picture didn't make sense, since there's no shoulder to pull over onto and the hill is very steep.

Betsy inhaled quickly.

Even though the other car was on the road, there was ample room in our lane to pass it safely. I tapped my brake pedal gently. When my foot came off the pedal, the brakes—much vaunted by the salesman as antilock brakes—remained clenched.

I turned the wheel gently to the right so that our path would veer away from the other car, but the brakes were still locked, so the wheels didn't respond. In less than a heartbeat we were in a skid. And it wasn't one of the fun kind.

The vision before me suddenly came into very sharp focus. We were perhaps 150 feet from the other car. We were traveling at about twenty-five miles an hour, so for three or four seconds we could clearly see what was going to happen. We even had a conversation about it

"Uh-oh," I said with a growing sense of dread. We were headed directly for the other car.

"Are we—" Betsy started to ask.

I said, "Hold on!" but Betsy didn't hear it.

We struck the other car almost head-on. I remember no sound, but the air bag filled the space in front of me instantly. The bag grew flaccid almost as quickly as it had inflated, and as it shrunk away, I saw the car we had struck rolling backward down the hill. Its rear end suddenly dipped as it went into a deep culvert, and with a sickening bounce, it came to a stop at a rakish angle, half in a ditch, half on the road.

Inside our car, the air seemed suddenly smoky. Even before asking Betsy whether she was all right, I heard myself saying, "Electrical fire. We've got to get out."

Betsy's first response was one word: "Elizabeth."

I released my seat belt and turned to the backseat. Elizabeth had started to cry, but like Betsy and me, she had been well protected by the intact passenger compartment. "She's okay," I reassured Betsy, unbuckling the harness that held Elizabeth. "But we've gotta get out of this car."

Betsy's responses are usually more methodical than mine—I seem capable of quicker and longer leaps, but she's more likely to land in the right spot a moment later. In this case, she returned her attention to the car.

As I hefted Elizabeth into the front seat, Betsy said, "It's dust."

"What is?"

"It's not smoke. It's dust," she said again, sniffing. "From the air bag?"

We got out of the car quickly. Once outside, I handed Elizabeth to Betsy.

There seemed no particular reason to hurry once we were clear of the car. There was no smoke outside. The fine mist we had seen was indeed a lubricant for the air bag, which had protected me from injury. Betsy, who had only her seat belt for protection, said she was fine, though her shoulder stung where the seat belt had gripped her. Elizabeth stopped crying almost as soon as she was in her mother's arms.

The car wasn't fine. The engine had stalled, and the vehicle's nose was angled to the edge of the road, one wheel over the shoulder. The driver's side front quarter was crunched, the body panels accordioned by the impact of striking the other car. The grille and lights were smashed, the hood sprung upward. Even in the dark, it was clear the car couldn't be driven. With a minimum of conversation, we continued down the hill on foot.

We were on our way to Charlie's house.

THE WALKING WAS LESS treacherous than the driving, so we were knocking on Charlie Briggs's door five minutes later. Through a window, I saw him rise from his chair next to the woodstove. His stooped frame was unmistakable.

He registered no surprise as he opened the door and immediately invited us in. His wife's voice could be heard down the hall. "Who's at the door, Briggs?"

"The Howards," he called over his shoulder.

He turned back and, uncharacteristically, looked me directly in the eye. "Bad night for driving," he remarked.

I laughed at that. The sound I heard myself making echoed like a cackle—it was more an involuntary discharge of nervous energy than an expression of amusement. "We've had an accident, Charlie. But," I added hurriedly, "no one's hurt. I need to call the police and get the car towed."

Charlie nodded, both in acknowledgment and as a gesture to the phone that sat on a desk a few feet away. The phone was a black desk model with a rotary dial. My hands shook slightly as I placed the calls.

While I was on the phone, Charlie's wife, Helen, appeared. She was dressed in an apron and immediately sat down with Betsy. Judging from their conversation, neither wanted to talk about the accident.

In a matter of minutes, arrangements had been made to meet a police car and a tow truck back at the scene, and for Diane at Long Barn Daycare to take care of Sarah for the night.

As I got off the phone, Charlie appeared with his coat and snow boots on.

"Come on," he said. "We'll take my truck."

I started to object but he already had the door open. The blast of cold, wet air that entered made me suddenly appreciative of his offer of a ride and his company. Betsy and Elizabeth remained behind with Helen in the warm house.

THE POLICEMAN ARRIVED FIRST. In less than five minutes, he had recorded the information he needed to file his report. He remarked as how he had gotten into a skid coming over. "Very slippery out there," he said. "Coulda happened to anyone." He patted the crunched front end of our car as he spoke.

The tow truck arrived and winched the car onto its flatbed. Another truck carried off the other car. Betsy appeared, too, on foot, with Elizabeth in her arms. I took my turn holding her while the cop told me to call his barracks in the morning to get the report number for my insurance claim. Charlie offered to drive us up the hill to our cottage, but I assured him we had had enough accidents on that hill for one night. I thanked him and he went home.

So did we. Only we decided to stop along the way at our new house.

THE SNOW WAS STILL falling. There was no wind, so the flakes seemed to be floating in slow motion before settling quietly on the ground. Only when we were within a few yards of the house did it become visible, rearing up out of the gray and snowy light like a fifties monster looming up on an old black-and-white television screen.

The house had no steps leading to its doors, just a rickety lawn chair positioned outside the back door. Betsy clambered up the equivalent of two tall steps—onto the chair, then to the floor of the house—and then I handed her Elizabeth and climbed inside, too.

A whiff of wood smoke greeted us. I fumbled in the dark to find the flashlight we kept by the door, then led the way. A pile of lumber filled most of the long hall. We had to squeeze between the bare studs of the wall and a two-foot-tall pile of wallboard to turn into the living room.

Betsy's face lit up—she had not seen the fireplace in use. I fed the coals some paper and kindling, and the fire quickly roared, the flickering flames splashing light around the room. This was no commodious living room, at least not yet. It was a construction site, but for those of us who had dreamt up this house, it was a promise partly kept. We were safe inside, the heat of the fire warming our faces. A spool of electrical cable and another broken-down yard chair provided seats. The three of us watched the fire together, and after a few minutes Elizabeth fell fast asleep in Betsy's arms.

I stood up and walked over to the wooden keg of antique nails. Earlier, when I was cleaning up, I had found that bottle of tomato wine Wilho had given me. It had been wrapped in a couple of rags. I took the bottle out and rummaged around to find two slightly used paper coffee cups. I rinsed them with water, opened the screw-top bottle of wine, and poured a couple of ounces into each cup.

I made a toast—"To our house and our health." Then we tasted the wine and it was—well, it was homemade, had been improperly stored, and might not actually have been good at all for anyone's health. We laughed and Elizabeth stirred briefly, blinking and looking around.

In the moment of quiet that followed, there was a noticeable creaking sound. We had been talking nonstop, half excited at the first fire in our house, half relieved after our close call in the car. The crackling of the fire had quieted, and on first hearing, the new sound seemed like it might have been the wind. On hearing it again, I knew it was a complaint of an unmistakably architectural kind.

Before I could say anything, Betsy raised her index finger and cocked her head. She was listening, too, having noticed the same *creeeeeaak* from above us. The creak was heard again. It wasn't a cracking sound; nor was it a crunch.

"It sounds like a sigh," Betsy observed.

"I'm going upstairs to look."

A ladder stood in the second-floor opening to which the stairs would soon rise. As I climbed, there was a creak directly above me. Once on the second floor, I continued to the attic opening and climbed yet another ladder. Standing on the second or third rung from the top, I shone my flashlight around the low-ceilinged space—there was little there but the chimney, along with two framed openings, one at each gable end. I noticed they weren't sealed with plastic; I had forgotten them in the rush of the afternoon. A small amount of snow was accumulating on the attic floor. *Tomorrow,* I resolved.

A memory of Tor House flashed through my mind. Only yards from the Pacific, Jeffers's house had lost its roof in a powerful storm one winter. But the noise interrupted my thoughts, this time louder. I pointed the flashlight into the corners, then up to the peak, swinging the bright beam along the length of the ridge. No obvious explanation for the creaking was to be seen. The building looked sound and true. I dismissed the Tor House comparison. This roof wasn't going anywhere.

A few minutes later, we abandoned our tomato wine and returned to our finished house, our old cottage, and waited for the snow to stop. Betsy called Diane and talked with Sarah. Her night at Long Barn Daycare would be her first away from us. She was happy there; Diane had managed to make her feel like it was a special occasion. I was the one who had trouble sleeping, my mind alternately replaying our accident and wondering at the strange sighs in our new house.

THE BUILDING WAS, OF course, fine in the morning. The creaking sound never recurred. When I asked a carpenter friend if he had ever heard such a noise, he shrugged. "Always happens. The parts are just, uh, getting to know each other. Like your butt and a new leather bicycle seat."

Some months later I found a preferable description in reading about Frank Lloyd Wright's Fallingwater.

At the time of construction, there was a fear that the structure would fail because its large cantilevered terraces reached out into space like immense diving boards, with no evident support below them. Engineers hired by the client had reviewed the working drawings. The calculations they did in their structural analysis predicted that the cantilevers would prove unstable, that they would deflect or "flucture." In a word? *Crash!* But when the concrete forms and the superstructure around the building were removed, the building didn't fall. Despite a continuing (and serious) problem with cracks in the concrete, Fallingwater still stands.

Our house tested no limits as Wright's did—ours was mere arithmetic in comparison to his calculus. Fallingwater is perhaps the most admired house of the century; our home would be nothing more than a vernacular exercise in a traditional style. Yet Fallingwater and our house each have a place on a larger continuum, and they aren't quite separable. Both reflect a mix of aspiration and trepidation that is common to house builders of all times and talents.

A game-show host might call that snowy night the moment of truth. But architects, with the understatement typical of the profession, use the term *negative moment* to describe the fear of failure at Fallingwater. In some sense, our anxious evening sixty years later was our negative moment. But our building didn't fall and we weren't hurt in our car accident. Ralph's (and Rumford's) fireplace functioned perfectly. Charlie and Helen Briggs were home to greet, help, and support us. We had Wilho's bottle of wine with which to toast our progress. That eventful night, with its accident and the creaking house, stands in my memory at the fulcrum of the building process.

A Winter's Work

Aim high in steering.
—Mr. Plumber

The construction site was silent. Back in my office, I sat before my computer, the desk almost invisible beneath strata of books, magazines, and catalogs devoted to tools. The plan was to finish the book quickly in order to get back to work on the house, but my notions of what can be accomplished in a given period of time are often a little optimistic. I like to assign some of the responsibility for that overreaching to Mr. Plumber.

Mr. Plumber was a driver's education teacher. My classmates and I got great glee from his name, but the man himself wasn't droll at all. I remember him as a gray and featureless man who wore spectacles and a nondescript wool sport coat.

His name might well have been lost to me altogether except that he drilled into his students one memorable phrase. He probably didn't regard his much-repeated advisory as words to live by, but it was a good deal more than just sound advice for driving on a highway.

"Aim high in steering," Mr. Plumber told us. "Don't look at the road immediately in front of you, but keep your line of sight higher." In practice, aiming high meant his student drivers were learning to drive a steadier path, making fewer steering corrections. Aiming high is a cardinal rule of safe driving, but it's also an approach to life. I like to aim high in steering, setting as a goal, for example, finishing the tool book in a couple of months, or building us a house, or building that house in—well, how long *would* it take?

That was a question to which there was no known answer. We had fixed one interim goal in our minds: to get the box closed by winter. That had been accomplished. But when would the floors be ready for sanding? What day would the plastering begin? When would we be able to move in? Since the book emerged more slowly than expected, setting a firm construction schedule was impossible.

Thoughts of construction distracted me from my writing. The HVAC crew was due in early January, but as December wore on, the work site remained quiet. That no progress was being made weighed on me like an overdue debt to a friend.

What was needed was a job to do that would provide a large sense of satisfaction without taking a great deal of time. One afternoon, pacing about my office, I glanced out my window. The electric blue of a tarpaulin caught my eye. Beneath it was the answer, an antique set of stairs I had been eagerly looking forward to installing.

SHORTLY BEFORE WE BEGAN to build our house, Don, the fellow who had recommended Ralph, called. That was when the design was slowly developing.

"You looking for a staircase?"

The question took me by surprise.

I hesitated before answering, "Well, yeah. I guess we are."

Don is usually a bit impatient, so the subtext *(Are we really looking for a staircase?)* was lost on him. He pushed on. "There's a house up here, an 1870s parsonage. You want the stairs before they tear the place down?"

All I could come up with was a question. "Maybe we'd better take a look?"

"Sure," replied Don, his tone matter-of-fact. "I've got a name and a number to call. You got a pencil?"

I took the information and hung up.

When Betsy heard about the staircase, she was immediately interested. "We want a staircase, don't we?" she said quickly. After a moment she added, "But I guess I'd like to have a look first."

THE PARSONAGE SAT ON a well-traveled thoroughfare on the outskirts of Troy, New York. Troy had had a great nineteenth century. Factories had produced quantities of stoves and other iron goods. A booming shirt-making industry had also produced a nickname, the Collar City. The twentieth century hadn't been so kind to Troy, and its surviving buildings were vestigial reminders that the city had once flourished.

Its congregation still maintained the Brunswick Methodist Church, but the parsonage had stood empty for several years, its façade overlooking the road that had once been the principal route from Boston to Albany. The current pastor and his family resided in a prefabricated home set well behind the church, a newer building that had more modern conveniences and was less expensive to heat. The old parsonage had been built just after the Civil War, during the prosperous time when the Troy Haymakers had been one of the original teams in the old National Association, the first professional baseball league and predecessor of the National League. The Haymakers were long gone, and soon this expendable house would be, too.

Betsy and I entered by a rear door, led by a church trustee. A soft-spoken man in his fifties, our guide acknowledged the state of the house—it was a mess. "But we are going to tear the place down pretty soon," he added apologetically. Asked when the demolition would take place, he allowed that no definite date had been established. He pointed in the direction of the staircase—"That's the parlor; the entrance hall and stairs are just to the right"—then left us to look.

The stairs were covered by a threadbare brown runner. The staircase didn't call attention to itself, but I noticed immediately that the railing was made of walnut, its tight grain a deep, rich brown. Many old balustrades have been painted, obscuring the cherry, mahogany, or other wood, but this appeared to have its original natural finish.

The backbone of the staircase, the stairs themselves, together with their wooden frame, or carriages, had been painted. The material was pine or some other softwood, and I suspected that the treads beneath the carpet were badly worn. But the whole effect was of a generously made and unspoiled staircase. Given its date, we knew it was constructed entirely of machine-made parts. But the feel was right: this was an old staircase, and while in very good condition, it bore the dents and scars and wear marks of a well-used piece of antique furniture.

I made a couple of quick sketches, then called out measurements to Betsy, who annotated the drawings with them. There were fifteen treads, including the one that was flush to the floorboards on the second floor. The railing began at a turned newel post and angled up the stairs. It then leveled off before making a 180-degree U-turn and continuing horizontally another nine feet, where it butted into a wall. The second-floor railing formed a balcony, overlooking the stairwell from above. This staircase looked very promising.

We were beginning to get cold when our guide returned. When asked whether there was heat or electricity in the building, he shook his head. He warned us that if power tools or lights would be required to remove the stair, we should bring an extension cord long enough to reach the basement of the church next door.

I asked if they had decided upon a price. "Whatever you think would be fair," was his simple answer. I told him we needed to go home and think about it, but that we would probably buy the staircase. He looked pleased.

As we prepared to leave, the abandoned house seemed suddenly colder. Betsy and I were both experiencing the same stiff-fingered, frigid-footed cold you encounter when you're waiting on a street corner on a wintry day. But the chill of an empty house is more profound. As we walked back through the house, there was evidence of the last inhabitants. Stacks of newspapers and a few half-filled boxes were littered here and there, left behind on moving day. A big pile of miscellaneous rubbish had been consolidated in one corner. All the closet doors seemed to be open, and the closet bars had hangers on them, a few still laden with unwanted articles of clothing. A glimpse into a laundry room revealed a cardboard box of detergent whose bottom had once been damp and then had blistered. Curds of soap surrounded it.

There was a pervasive sense of sadness about that empty house. But its staircase would soon have a new home.

WHILE AT MY DESK musing on going back to work on the house, I called Mike to ask him if he wanted work the following Saturday. My plan was to stay at my computer from Monday through Friday, but to spend weekends working on the house. Mike and I agreed to start at eight o'clock, and on the appointed morning he arrived at the cottage promptly, driving his compact pickup.

The dismantled parsonage stairs had gone in two directions. With the help of a friend, I had dismantled the railing, balusters, and trim pieces and stowed them out of the weather in his barn. But the rest of the staircase—the painted treads, risers, and carriages—sat in the same copse of evergreens near our old house where the salvaged doors and windows were stored.

Mike parked his truck as close to the salvaged goods as possible. We peeled the woven nylon fabric off the stacks. A family of mice that had established a nest beneath the tarp scampered off. A quick inspection indicated the doors, windows, and stairs were cold and damp but unharmed. The early-nineteenth-century six-panel doors and six-light sash, along with the mantel Ralph had used, had come from a salvage dealer in Connecticut, all for the price of three thousand dollars.

We carefully packed the windows upright in the bed of Mike's pickup. The cargo space wasn't large enough to take all the sash in one load, since there were about sixty in all. It required several trips to move the windows and the thirty-odd doors. We stood the doors up in the main hall, while the windows went in stacks in another area where, on another day, we could sort them by size and condition.

The staircase was both easier and harder to move: only one trip to the house would be required, but the stair structure was fourteen feet long and awkward and weighed perhaps two hundred pounds. We scratched our heads about how best to muscle it onto the roof rack of my van, but we soon managed to get one end up, then swung the other on. We secured it with clothesline, then drove carefully up the hill. It took us almost two hours to get all the salvaged parts to the house.

We took a coffee break and examined the three pencil sketches of the stairs in their original site. One was an elevation, with the stairs coded by numbers 1 through 14 and the balusters on the stairs represented by letters (1A, 1B, 2A, 2B, and so forth). A second drawing was a plan that showed the rough opening in the second floor (it was 104½ inches long and 46⅜ inches wide) and also laid out the exact positions for each of the trim pieces that would finish off the opening. A third sheet had several smaller sketches showing floor-to-floor height, the rise and run of the stairs (8-inch rise, 10¾-inch run), and other details. As we looked at each drawing, we talked the process through. Rather, I talked a lot and Mike interjected occasionally.

Building bears a distant resemblance to chess. Failing to think a couple of moves ahead is the quickest way to put yourself in jeopardy. We were dealing with a stair structure that was awkward and heavy enough to be potentially dangerous. I'm also a big believer in doing things once, especially when the work involves hefting hundreds of pounds.

This staircase had been raised many times in my mind, so I knew we couldn't just lift the thing up and expect it to pop into place like some precisely machined replacement part. This was a staircase that had been fitted by hand in a house utterly different from its new home. It would have to be made square, its treads level and its risers plumb. The carriage that fastened to the wall would have to be set flush so the trim board and drywall that would eventually butt against it would be at the proper depth.

The first step would be to find the point on the first floor that would be precisely below the top end of the stair. To do that, we dropped a plumb bob, a pointed weight attached to the end of a line. The bob swung like a pendulum for a few seconds, but when it stopped, the line was vertical and the tip of the bob indicated the point on the first floor precisely below the fixing point above. Using the measurements taken at the old parsonage, we then measured off 140 inches, pinpointing exactly where the first riser would be set. That was important because we then nailed a pair of boards, called ledgers, to the floor to hold the stair in place.

Imagine a book leaning at a steep angle against a wall. Given the pull of gravity, the book wants to slide down, returning to horizontal. But if you put another book flat on the floor to brace it, the angled book will stay in place. A staircase is installed the same way, with boards nailed to the floor to "foot" the staircase base.

We eyeballed the adjacent stud wall. One stud had to be replaced, since it had twisted and protruded into the space the stairs would occupy. Then we were ready to raise the stair. With a few grunts and groans, we lifted and slid it into place, grateful when we leaned it against the header of the opening above and the structure of the house took the weight off our shoulders. We fastened it temporarily, driving three-inch drywall screws through the carriage into the adjacent stud wall. Then we got the considerable satisfaction of climbing to the second floor for the first time on a real stairway.

Once upstairs, we leveled the top stair, using a spirit level as a guide and cedar shingles as shims. The shingles, which taper in thickness from about three-eighths of an inch at one end to less than one-eighth of an inch at the other, make perfect spacers. Next we checked the bottom tread—it was level. Eventually there would be a stud wall with a plaster surface beneath the open side of the stair, but temporarily we added a couple of braces to keep it steady.

We were finished just in time for lunch. It had been a good morning's work. The stairs would make moving materials as well as ourselves to the second floor much safer and more convenient. And I had now worked with Mike enough to know that he was the answer to the who-will-wear-Mark's-shoes question.

Before we went off for a sandwich, I fastened the pencil sketches to a nearby stud using a staple gun. We would need those drawings when installing the newel post and railing, but that lovely old walnut was too precious to have in place during construction. The balustrade would come later.

IN THE WEEKS THAT followed, Mike arrived every Saturday and Sunday. When he got laid off from his landscaping job, just after Christmas, he asked if he might work for me during the week, too. I agreed immediately.

Though I had to remain at my computer Monday through Friday, there was ample work for him to do. The challenge was to plan the job so that Mike could proceed by himself without having me looking over his shoulder. That required an adjustment for both of us.

We started with framing. While the outside walls had been framed, as had the essential bearing walls inside, which carry the weight of the structure, many partitions still needed to be built. The second floor, for example, was an open space without the walls that would define three bedrooms, two baths, and nearly a dozen closets, cabinets, and built-ins.

Each day, Mike and I would begin with a session of layout. We would mark the exact location of a partition wall on the plywood subfloor using a deceptively simple device called a chalk line. We would stretch the line taut between two points on the floor, and then one of us would lift and release it. The snapping of the string would deposit a line of chalk on the plywood. I would crosshatch the chalk line with pencil marks for the location of corner posts, studs, door openings, and other elements.

After I had gone back to my writing, Mike would cut the studs and assemble and raise the walls. Most weren't large, and one man could manage the job. At coffee breaks and lunchtime, he usually had a few questions. Some days, after an especially productive morning working at my computer, I would work with Mike in the afternoon.

By the second week in January, the interior partitions had been framed. That meant the wiring could begin. Given my work as an electrician's assistant years earlier, the wiring at our house seemed like the easiest part of the task. Again, I would lay the job out (none of the electrics were on paper, so we made the wiring scheme up as we went along). Then Mike would put it into effect.

First, the boxes for switches and plugs were nailed to the studs. Next, wires were run connecting the boxes to the basement, where Mark and I had spent a rainy day installing the main panel back in August. Mike would drill holes through the wall studs and floor joists, then snake the plastic-sheathed cable into the boxes, leaving about eight inches of wire protruding from each one.

The electrical code specified the number of plugs or lights per line, so in most cases, a second wire would exit the box, bound for the next box on the line. None of the wires would be connected to plugs or switches at this stage, nor to the breakers in the panel. That task would come later; this was just the electrical rough-in. The size of the cable was also specified by code —twelve gauge for the plugs, fourteen for the lights. That meant the plugs would have twenty-ampere breakers, suitable for appliances, while the lights would have fifteen amps. The cable in most cases held three wires or conductors, although for three-way and four-way switches Mike ran four or even five conductors.

He did the wiring well, having himself been an electrician's assistant in the past. But part of me wanted to be there, too, doing it with him, even though it was a bitterly cold work site, since the house was still unheated except for an occasional fire in the fireplace. In fact, that January was becoming the coldest one most people could recall. Even Charlie Briggs found his recollections challenged by numbers like twenty-five degrees *below* zero. He shook his head when asked if he remembered a colder winter. He couldn't. The HVAC guys couldn't have arrived at a better time.

The ground outside was frozen four feet deep. We know that for a fact because a twentyish rookie laborer working for the heating contractor began the job by jackhammering through the tundra. He was running the fill pipe for the oil tank, and installing it proved to be a full day of backbreaking work. Presumably he was at least warmed by his exertions.

The heating crew brought a red, submarine-shaped device called a salamander. It was a portable heater, fueled with kerosene, and they set it up in the cellar. A built-in fan then drove the heat from the burner into the surrounding space. In a matter of minutes, the basement became the warmest place in the house.

Mike watched the HVAC crew muscle the 350-gallon oil tank down the cellar stairs. When it came to getting the furnace to the basement, they drafted him to help. One man did the furnace installation, while another worked on the delivery system, a network of sheet-metal and fiberglass ducts that would eventually bring heat to registers in the living spaces. It would take the crew about a week to finish.

Meanwhile, Mike was completing the electrical rough-in. Once it was done, the insulation could be installed. The timing was impeccable, if accidental. As the heating system neared completion, Mike began stuffing batts of fluffy fiberglass into the bays between the bare framing members on the exterior walls.

This was one job I didn't really want to help Mike perform. Fiberglass, as its name suggests, is made from glass, fine strands that are spun into thick blankets. The material itself doesn't keep the cold out. As Count Rumford pointed out two hundred years ago, the air trapped in the fibers does the insulating.

The installation didn't take a great deal of skill. The batts were cut at the factory to standard widths that fit between joists and studs set sixteen inches on center. Some pieces of insulation needed to be cut to length, but a utility knife did the job easily. The tolerances weren't high—the trick was to cut each piece an inch or two long, and the excess would help snug the piece in place. Then the paper lining of each batt was stapled to the wooden frame. Except for cutting and fitting around electrical boxes, that was about all there was to it.

The bad news was that some of the fine strands of glass would inevitably get airborne. Thus, wearing a mask was essential. Without one, a tickle would turn into a cough and you wouldn't want to think about how many tiny shards of fiberglass you had lodged in your lungs. While the fibers wouldn't actually cut the skin, they would produce itching, so gloves and long sleeves were also necessary. Goggles, too, were a good idea. The insulation installer ends up looking like an arctic explorer. Which wasn't such a bad thing for Mike in that cold house.

Mike finished the job in a matter of days and didn't mumble a word of complaint.

We spent a Sunday covering the insulation with polyethylene film. Managing twenty-five-foot-long sheets was much easier for two men than for one. The poly came in rolls, and we stapled it in place, trimmed off any excess, and sealed the joints with tape. This would create a vapor barrier to prevent moisture inside the house from traveling outward and condensing in the wall cavities, as well as to limit the infiltration of cold air from the outside.

I wanted to continue working at the house, but the writing was going more slowly than expected, so during the week I remained at my desk. On the other hand, Mike was making surprisingly good progress and I could take some satisfaction in that.

One noontime I arrived at the construction site, only to find the HVAC crew had gone.

"Gone for lunch?" I asked Mike.

"No, they're done," he said.

I listened. Sure enough, I could hear the furnace running. The place still looked like a work site, but with a heating system, it had one of the comforts of home. In that moment, the realization struck that my frustration wasn't entirely about not being able to work on the house. The vision of a place where we could go to live was beginning to crystallize. I was beginning to want that very much.

MIKE'S NEXT TASK WOULD be to scrape the antique doors. The job had to be done, but it was unpleasant work. I almost apologized to Mike in describing the job to him.

The time spent sitting beneath a tarp had helped loosen what, on some of the twenty-six doors we had selected, were ten or more coats of paint. The rising damp from the decaying leaves and pine needles had caused some of the paint to alligator, readying the surface for scraping. Those doors could be scraped mechanically, but others would be more difficult. On them Mike would use a heat gun and a heat plate to get the paint bubbling hot before using a scraper or shave hook to remove it.

Even with good tools, it wasn't easy. Occasionally the layers of paint on a door will almost crumble off; more often, the work progresses an inch or two at a time. Some areas are more difficult than others. It takes patience as well as elbow grease. We had to take care when using the sharp tools, too, since the doors were all of pine, and the soft grain was easily gouged.

I took a few turns at scraping, but Mike proved more adept. He didn't hate the work as much as I would have: there's some satisfaction in controlling a small job like a door. You put it on the scraping platform, you scrape one side, you scrape the other. You do a bit of touch-up with a sander. Then you move the door from the unfinished stack to the finished one. There's a sense of closure about it. My approach was to scrape about a tenth of one side, then look up for a moment only to see an imposing stack of doors still waiting for me. Mike was better equipped for the job than I was.

The living room had become a scraping shop, with doors, scraped and unscraped, flanking the area where Mike worked. On the floor, a thick, gritty blanket of paint scrapings accumulated. A neat worker, Mike periodically swept the dust into a growing mound in an out-of-the-way corner of the room.

On most weekends, we worked on the windows.

We had decided to use old windows. This may seem sacrilegious in an age where, almost single-handedly, the replacement-window business seems to underwrite late-night television. But old windows are different.

The glass is wavy. The surface undulates, the thickness varies. The light is refracted and reflected differently, too. To watch a candle burn, its light flickering and dancing, through a window sash of old glass is to feel you've just been offered a fleeting glimpse through time.

Some of this is in the eyes of the beholder, but Betsy and I wanted old glass. And old sash, too, because the strips of wood between the panes, the muntins, are thinner. In the 1830s, when most of our window sash were made, builders wanted the wooden elements to disappear; they even painted the sash black so that in the light of day it would be virtually invisible.

I had sorted through the windows, making stacks of the various sizes. Few were identical, since they had come from perhaps ten different houses. Many were similar in size, though, about thirty inches wide and twenty-eight high, with panes of glass that were nine by twelve inches. There were lots of broken panes of glass, and the putty that held the individual pieces in place was badly deteriorated, cracked, crumbling, or, in many cases, missing altogether. Much restoration work would be required, but at least the sash existed.

The window frames did not: those we would have to make.

That may sound difficult, but once I've dismantled something, whatever mysteries it once held for me usually vanish. Replicating what has been taken apart isn't a given, but with a window, the elements were surprisingly simple. In remodeling our old cottage, the windows hadn't been old enough to be interesting (most of them dated from 1941), but in removing them, I had taken several apart, piece by piece, to see what the constituent parts were and how they fit together.

They were double-hung windows, the kind that have two sash that slide up and down within the plane of the wall. The frames consisted of jambs and a sill and were effectively simple wooden boxes without tops or bottoms. The top and side jambs were at right angles to one another, while the sills had a slight pitch (typically about three degrees) in order to shed rainwater. The exterior trim boards were nailed directly to each box on the outside, and the unit was then fastened in place by nailing the exterior trim to the wall of the house. Once the window was in place, the interior trim could be added later.

The sash moved within the frame in channels that were defined on the inside by thin boards called stops and a narrow strip between the two sash called a parting bead. Though designs vary, in the windows we planned to make, the exterior would have no separate stop; rather, the exterior trim would extend inside the box, forming a lip that would serve as the outside stop.

The various parts of a properly manufactured window frame aren't simply butted and nailed together. Other kinds of joints are required. For example, the head jamb and the sill have to be let into the side jambs, meaning that grooves called dadoes must be cut into each jamb near its top and bottom. After the application of liberal quantities of waterproof glue, the head jamb and sill are then slid into the grooves. The parting bead also sets in a dado, which runs the length of the jamb; it's nailed but not glued, to enable the sash to be removed later.

No single step in manufacturing a window was particularly difficult. The challenge was that there were four pieces to the basic frame, four to the exterior casing, three lengths of parting bead, two inside stops, and a range of interior trim pieces. The sill alone, because it had several bevel cuts, required roughly ten different saw cuts. And then there was the hard part: well, not for everyone, but for us.

We wanted our house to have some of the qualities of an older house, and the exterior trim distinguishes the well-built early American house. Today, builders and homeowners usually content themselves with very plain window casings of flat boards. We wanted something more decorative, but we didn't quite know what. As we examined the stacks of windows, I decided to take a short ride to visit our front door.

MORE THAN A YEAR earlier, a telephone tip had led Betsy and me to a shed on a backstreet in Hudson, New York. A onetime whaling town on the Hudson River, the city had been established in 1783 by sailors from Nantucket looking for safe harbor from British ships. Though it had a prosperous-looking main street lined with antique shops, Hudson was a city with more of a past than a present.

The antiques dealer we met was known to us by sight. Like us, he frequented country auctions, and at one such event, he had bought a doorway from a once-grand house that had been abandoned. When he showed us the doorway sitting on the dirt floor of his rotting carriage shed, we knew we had to have it.

There was more than just a beautiful seven-foot-tall door with eight raised panels. There were pairs of fluted pilasters, two on each side of the doorway on the inside, with two more pairs on the exterior. The pilasters framed sidelights. The sidelights were four-foot-tall, windowed openings, with bentwood tracery in a repeating figure-eight pattern on each of the four-by-nine-inch panes of glass. At the crossing points, I could just detect rosettes beneath the dense layers of paint. They were made of lead that had once been gilded. The woodwork, which had panels, beads, capitals, cornices (inside and out), and other details, was finer than any work I could aspire to. If we were to commission a new door of this quality, it would cost five thousand dollars or more.

On the spot, we had bought it for fifteen hundred dollars. With a little help from two sturdy friends and a borrowed pickup, we stowed the doorway in a disused barn on a neighbor's property.

Mike and I went to look at the door for inspiration. We would be restoring twenty-two sets of sash to make twenty-two double-hung windows, so we needed to decide how we would finish off the exterior of the windows. The four pilasters on the doorway provided the answer: we could have a pilaster on either side of each window, complete with base and capital, that supported a cornice above. I measured the pilasters on the doorway; they were about five inches wide. We thought out loud a bit: the doorway was a much larger unit, but why couldn't we adapt the same details on a smaller scale? Standard stock comes in nominal four-inch widths (after planing, that translates to three and a half inches). That seemed about right for smaller windows that would flank the larger doorway.

Using standard materials always saves time in the fabrication process—it's one less rip cut on the table saw to narrow the stock to the size you want, one less pass over the jointer, the tool that planes off the saw marks. Operating under the same principle, I decided to make the basic casing out of nominal one-by-six stock (actual dimension, three-quarters of an inch by five and a half inches). Mike suggested that the pilasters, which would be applied on top of the casings, might be made of what's called five-quarter, meaning wood that is nominally one and a quarter inches thick. The added thickness would provide more stock for cutting the grooves, or flutes, in it and would add more relief, making the pilaster stand out more prominently. With a headpiece across the top, each window would take on the look of the doorway.

We set about making a model. Even when a production run will consist of fewer than two dozen finished products, a model is essential. I referred to a couple of books on my shelves for guidance, and then Mike and I went to work in the makeshift workshop we had set up.

We started by making the basic box. That meant cutting the side and head jambs. First the raw stock had to be ripped to the width of the openings the jambs would fill. Next the pieces had to be cut to length so that they would fit loosely inside the rough openings.

The side jambs then had to have grooves cut in them to fit the sill and head jamb; these grooves were to be located so that the distance between the inside of the top and bottom of the frame suited the window sash exactly. We tacked the pieces together and set the assembly horizontally on a workbench. It took some cutting and fitting to get it right. We had to remake one jamb because the left and right jambs aren't identical: since the sill is joined at an angle, the left and right sides are mirror images of each other.

In two hours, we had a box. The process had required setting and resetting the table saw twice for ripping and three times more to accommodate the dado blades we needed to cut the grooves (one for the head jamb, which was made of boards that were three-quarters of an inch thick; one for the sill, which was made of five-quarter; and one along the length for the parting bead, which would be a half-inch thick. We used the radial arm saw to cut the boards to length. A band saw was required to make the L-shaped cuts that formed "horns" on the sill, the portions that extend beyond the width of the frame and against which the casing would eventually butt.

This was our first chance to work together in the shop. We didn't talk a great deal—Mike's a man of few words, and workshop conversations are necessarily piecemeal, given the frequent interruptions of the loud saws. He was very different from Mark, who had desired to learn everything he could from every moment. While Mark was back in college immersed in the study of European history, Mike arrived on the work site every day dressed in camouflage, the same warm clothes he wore in the woods when hunting for deer, a rifle cradled in his arms.

Mike was a worker bee. He had little concern with the horizon but worked very hard for his paycheck. We found each other companionable, and like ballplayers, we went about learning each other's moves.

IN REMODELING OUR COTTAGE, I had made some moldings using antique hand planes. Bought at flea markets and antiques fairs, each of those planes had a wooden body with a metal plane iron that protruded from the bottom. The base of the plane and the iron had matching profiles, which were the reverse of the molding produced. Some hand planes are convex and some concave (called hollows and rounds, respectively); others shape complex curves, such as ogees and other molding profiles. Some molding planes have square profiles.

There's great pleasure to be gotten from working with such planes: there's a simple joy in seeing the shaving curl out of the plane's throat, hearing the satisfying, soft scrunch of the blade slicing into the grain, and then feeling the tautness that develops in the forearm muscles after working the plane for a few minutes.

The modern power tool that accomplishes the same purpose is the router. It is utterly different. Rather like an electric drill, the router has a motor-driven shaft with a chuck mounted on its end. The circular blades, or bits, are clamped in the chuck. Unlike an electric drill, the router has a base through which the bits protrude. The operator grips the router by its two handles, directing it to cut grooves, mortises, or molded edges.

Unlike the peaceable quiet of the molding plane, the router howls when at work, emitting a screech at an earsplitting pitch and volume. No sensible operator would ever use one without ear protection. But the router is a tool that makes sense in a production line: it'll cut twenty-five feet of molding in the time a skilled artisan can make five with a hand plane.

In the world of manufacturing—and we were establishing a small manufactory to produce our window units—efficiency has a lot to do with operations. Ripping a piece to the proper width is one operation; planing the edge smooth is another. Cutting it to the correct length is a third. Fluting the pilaster shafts required four passes with the router. The capitals consisted of three distinct pieces, each of which was ripped, cut off, and routed to a different size and shape. To mold the edges, the machine had to be stopped between operations and reset in order to shape the ogee molding that topped the capital; another bit had to be fitted for the astragal, or half-round molding, that sat atop the shaft as part of the capital.

The router was an invaluable tool that made making our elaborately decorated window frames practical.

MIKE CONTINUED WORK THROUGHOUT January and into February. Since there was only one or two of us at the job site, we would usually have lunch at the cottage. There Mike proved to be as adept at slicing bread and assembling a sandwich as he was at his construction tasks. He even looked the same. He wore a cap when doing carpentry, preparing food in the kitchen, and eating lunch at the dining table. I never saw him without it.

One February morning, Betsy had taken Elizabeth for her one-year-old well-baby checkup. When they came home, Mike was just finishing lunch. The moment Betsy appeared with Elizabeth on her hip, I could see something was wrong. Perhaps Mike saw it, too, as he quickly excused himself and headed back to the work site.

"Elizabeth has lead poisoning," Betsy blurted out after he left. Tears were welling up.

I knew that lead was toxic, especially to children.

Betsy was experiencing more than a panic about Elizabeth: there were symptoms of "I'm-a-bad-mother syndrome." In our three years of parenting, I had observed that when one of the girls fell ill, her adrenaline might rise, or on the other hand, self-doubt might surface and her spirits fall. Betsy showed signs of crumbling.

In our marriage, when one of us becomes alarmed or disheartened, the other is the outrigger, instinctively reaching out to stabilize the craft. This time I put an arm around Betsy and, out of instinct, looked to Elizabeth, who was engaged happily with the circle of toys around her. Betsy followed my gaze. This was not a child in dire straits. Involuntarily I smiled. Betsy visibly relaxed.

"What did the doctor say?"

She told me what had happened. The nurse had done the usual preliminaries, including weighing and measuring Elizabeth. The doctor then engaged Elizabeth in a game of peekaboo and asked Betsy a variety of questions. The nurse then gave Elizabeth a DPT (diphtheria, pertussis, and tetanus) shot. Finally, she did a prick test, in which she stuck a needle into Elizabeth's fingertip to get a drop or two of blood, which she immediately tested for lead.

"At a level of ten, they begin to get worried," Betsy said. "Hers was over *forty,*" she said with quiet emphasis.

Her worry made me somehow calmer; that seems to be part of the outrigger effect.

"They want a venous test done. It's not one they do in the office, so we need to go to the hospital this afternoon to get it done."

"Why another test?"

"The prick test is just to screen for high lead levels. It's probably not very accurate. Her level may be much lower. . . ." Her voice trailed off.

I realized that we knew little of Elizabeth's condition for certain, and that Betsy's focus was shifting from her worry to the action she needed to take. That's her nature: once a target for her energies has been identified, she aims for it, and the inertia is overcome.

Yet a question nagged at me. We had no lead paint in our cottage, where all the exposed surfaces had been painted by us with oil-based paint. Something was wrong with the picture: Where could the lead have come from?

Our old doors came to mind, but I dismissed them. That was at the work site, and Elizabeth was rarely there. It just didn't add up.

ELIZABETH'S SECOND BLOOD TEST came back much lower. The result was twelve micrograms of lead per deciliter of blood, and while that may not sound like much, the doctor advised that treatment was warranted. He prescribed an iron supplement.

Taken by mouth in a fluid suspension, the iron would bond to the lead in her system and she would excrete it.

When we heard the news, Betsy and I both felt better—but it was a qualified better. We felt like travelers who discover after a blowout that the spare tire is fully inflated. It's a great relief, but one that is followed by the realization that the journey will have to continue without the insurance of a spare. In our case, the nagging worry was that we still didn't know from where the toxic lead had come and, presumably, was still coming. Our confusion wasn't lessened by the test done to determine Sarah's lead levels. The finding was that hers were within tolerable limits.

By state law, the doctor was required to notify the board of health, so in a matter of days, our phone rang. Betsy answered a number of questions, one of which was whether we would like an on-site inspection to identify the source of the lead.

My first instinct was, *No thanks, we don't need a bureaucracy interfering with our lives*. Betsy's response was more measured. "She really sounded reasonable on the phone," Betsy said. Eventually, I climbed down off my don't-trust-the-authorities soapbox and relented.

A team of a nurse and a technician arrived. Betsy and I had agreed that we had nothing to hide and much to gain from solving the problem, so without hesitation we told the two visitors about our cottage home, its renovation, the new house under construction, and the patterns of our family life. The nurse asked questions about Elizabeth. The technician looked bored and said little. But the moment I said, "We're scraping the paint off some antique doors," his expression lit up like the indicator light on a metal detector. In minutes the technician was on his way to the work site with me as his guide.

That those doors had been painted with lead paint at some

time in the past, we took for granted. At my insistence, Mike habitually wore a respirator to protect him when scraping. But I had dismissed the notion that the scraping could have been poisoning Elizabeth because we had taken pains to be sure neither child had been anywhere near the scraping area. Since the weather had gotten cold, Elizabeth had visited the work site on Betsy's hip no more than a couple of times and had always been in either Betsy's or my arms. Were such fleeting visits enough to expose her to lead? The technician said probably not as we drove to the new house.

He took scrapings from the doors and the staircase. "We'll test these," he said. He seemed confident he had found the source. "We'll call you with the results next week."

Betsy took the follow-up call from the nurse. That evening she repeated for me what she had learned, and the picture became suddenly clear. They had found lead on the doors and the staircase. While that was expected, what was surprising was the means by which Elizabeth had probably ingested the lead.

The nurse explained that fine particles of lead adhere to fabric. When I scraped a door or even walked by as Mike was scraping, airborne lead dust would get on my clothing. Back at home, Elizabeth, always an affectionate child, would climb onto my lap. Some of the dust would then get on her hands and clothes. Like any one-year-old child, Elizabeth used her mouth to explore her world, so lead particles soon found their way into her system.

The next day, Mike and I fashioned a lead-containment room. We stapled sheets of polyethylene to the walls, floor, and ceiling of the ten-by-fifteen-foot space. All the antique doors were moved inside its confines. Mike agreed that he would do all the rest of the scraping himself and that he would designate a set of work clothes for use in the lead room. He would change into them upon entering and take them off upon leaving. Finally, we laboriously vacuumed every mote of dust we could find in the house.

As February became March, the windows began to take shape. They took time, since each window had so many parts. The frame alone required seven pieces, no two the same. The pair of pilasters accounted for sixteen parts. In total, each window consisted of thirty-three elements, not including either the sash or the trim on the interior; the latter we hadn't even milled yet, but I was guessing it would consist of approximately fifteen pieces. Sound complicated? Consider that cutting and planing the pieces required roughly 150 operations and multiple setups of the saws and router. We also had to cut extras of every piece because mistakes get made, pieces get cut too short, and splits and dry knots emerge where you hadn't anticipated them. Going back to refabricate pieces in ones and twos would have been very time-consuming.

The pieces did come together, and as the stack of assembled frames grew taller, Mike moved on to painting them. First a coat of primer was applied, then two coats of paint. Mike reputtied all the sash, then primed and painted those, too. Together we fitted them with state-of-the-art weather stripping.

How much did all this cost? The old sash had been $5 apiece; at two per window, that was an inexpensive $10 for each unit. Virtually all the wood that we used to fabricate the frame was number two pine, meaning it had knots in it and cost about $1 per square foot of one-inch stock (in carpenter's terms, that's a board foot). Each window required between thirty-five and forty board feet of pine. With nails, paint, and weather stripping added in, the cost of materials was a bit less than $50 per window.

Labor cost? My labor was free. Mike was working for $12 an hour. Each window required roughly a dozen man-hours, on average eight of Mike's. Mike's time cost me about $100 a window, so the out-of-pocket cost was roughly $150 per window for labor and materials. To have had windows milled with the same degree of detail would have cost at least two or three times as much.

More important, since the sash and the glass were antique, the windows looked old. These were Federal-style windows that bore a resemblance to those we knew from the streetscapes around us. Those windows, like the rouge on a clown's cheeks, pop out—significant details that would help locate our house on the time line of American architecture. That was, after all, part of what we were shooting for.

My work in the months of January and February had been limited, but the windows represented a major accomplishment. After the weeks invested in making them, the installation could have seemed anticlimactic. The simple process involved removing the poly that covered the openings—the weather and wind had rendered it translucent, a foggy gray. Then we set the windows into the openings. After leveling the bottom with shims, we nailed the exterior casing to the framing of the house. In the time it takes to eat an ice cream cone, we put a window unit in place and moved on to the next one.

Once the front windows were in position, I walked along the driveway away from the house. I didn't turn and look over my shoulder but walked purposefully to a distance of seventy-five feet or so and then turned around. I raised my gaze slowly, wanting to take in what I saw.

The sight made me smile. With the windows in, the place felt more like a real house than a dream.

The Common Hours

If one advances confidently in the direction of his dreams, and endeavors to live the life which he has imagined, he will meet with a success unexpected in common hours.
—Henry David Thoreau

Filing your taxes, taking your last exam, or completing a job of almost any kind is accompanied by a sense of liberation. The moment the manuscript I had been working on was completed and entrusted to the United States Postal Service, I felt like kicking up my heels. Instead, I buckled on my tool belt. The month was March, but with the manuscript on its way to the publisher, I was finally able to go back to work on the house in earnest.

The first job to do was the plumbing.

Different people are daunted by different aspects of construction. Often the electrical work seems the scariest. The fear may be of fire, electric shock, or even electrocution, but whichever the case, many novice homeowners are terrified of working with wiring. For me, plumbing was intimidating.

When we bought our cottage, more than a dozen years earlier, I hadn't the knowledge to identify serious problems in examining the structure and working systems. If I had, Betsy and I would have been shocked to find that essentially everything needed radical reworking. Our ignorance meant that we discovered the problems one at a time. After taking possession of the house, the plumbing was among the first.

The well went dry our third weekend living there. Within a few months, we realized that the kitchen was in the wrong place to suit our needs, lining the side of the living room as if the house were a tiny studio apartment. We wanted a larger kitchen, one discrete from the living room. The bathroom was actually in two places, with the tub in a virtual closet and the lavatory and sink two rooms away. At first we thought we could adapt, but the awkwardness of the arrangement just wouldn't go away, even though relocating the kitchen and bath meant replumbing the whole house.

The process began with a close inspection of what was there, flexible plastic pipe of the cheapest sort with hose clamps at the joints. None of the system was worth salvaging, but the overall scheme of things became evident in dismantling the pieces. The basic principles were that the supply pipes brought clean water to the fixtures, while the waste pipes took the dirty water away. The supply lines were small (less than an inch in diameter), the waste lines larger (about four inches). The water coming in was pressurized, which is why the water flowed when a tap was opened. The water and solid waste leaving the system was not pressurized but relied upon gravity to drain.

In a properly plumbed house, each fixture also has its own trap. This U-shaped space is built into every toilet or incorporated into pipes adjacent to the tub or sinks. The traps remain filled with water at all times, blocking sewer gases from entering the living spaces of the house. Instead, the gases escape through a vent pipe that rises through the roof. The vent also allows the waste to flow smoothly out (without venting, the waste pipes don't drain properly, like a water-filled straw when one end is sealed with a finger).

The plumbing seemed remarkably simple. The more difficult part proved to be replacing the old with the new, but that taught me an important lesson.

The job consumed several weekends, during which we had to rely on an old outhouse and containers of water carried from a nearby stream. A book on basic plumbing provided a working vocabulary for ordering parts at a plumbing supply store. It was at such a shop, on my third visit of the day, that my invaluable little insight was gained.

The goal was to get the kitchen sink functioning by dinnertime—I had promised Betsy. A 7:30 A.M. trip to the store had been followed by another at lunchtime for several forgotten fittings. At about 4:00, I was stuck again.

The supply lines were in place immediately beneath the sink. Aligned above was the base of the faucet with its two connections. Completion was very close yet seemed far away. The problem was how to connect the lines to the faucet. A standard threaded coupling wouldn't thread onto both at once—when I threaded the coupling onto the faucet, it unthreaded the connection to the supply line. No answer seemed evident on paging quickly through my plumbing book. I jumped in the car and raced to the store, hoping to get there before its 4:30 closing time.

I made it and was doubly lucky because the man on duty at the counter was Wendel, known universally as Wendy. He was a likable fellow and a weekend musician who played banjo in a local bar band. He had waited on me earlier in the day, and he looked a little surprised to see me again.

He smiled. "Not quite there yet, eh?"

I shook my head. "A fitting or two short, sad to say."

There were no other customers in the store. Wendy reached for the drawing he had spotted in my grimy hand. "Let me have a look."

I hesitated. "You might not be able to make much sense of that."

He looked over his reading glasses at me. "You might be surprised," he observed, then looked down again.

He absorbed it quickly. "I see what you want to do. So how far have you gotten?"

I explained the problem.

Without hesitation he said, "Why don't you try a union?"

What did that mean? He probably wasn't suggesting a call to the local plumber's union, since Wendy himself was a do-it-yourselfer on a limited budget. But that didn't make his meaning any clearer.

He pointed down the aisle at a honeycomb of bins. Like an old mechanical adding machine, my brain began to process the meaning of his gesture. The bins were filled with plumbing fittings, and yes, right, I remembered vaguely a reference in the book to something called a union.

Wendy must have heard my mental gears grinding.

"You don't know what a union is, do you?" he asked simply.

That was, in retrospect, a moment that looms large. Like most men, I occasionally suffer from "male answer syndrome." Ask the average man a question to which he does not know the answer, and he will use related knowledge to extrapolate a reasonable-sounding response. He may believe it to be true and will often defend the invented nonsense well beyond reason. That's part of the syndrome, too.

That afternoon, some mix of exhaustion and frustration led me to confess my ignorance. "No," I told Wendy, shaking my head. "I'm embarrassed to say I don't know what a union is."

"Look," he said, his voice firm. "I'm going to tell you two things. First, a union is an assembly you use to fasten fixed pipes, especially when they'll have to be disconnected periodically, like beneath your kitchen faucet. It'll solve your threading problem, too. Follow me and I'll show you where they are. You'll figure out how they work in about a minute.

"But you know what else? If you don't know how to do something, you gotta ask. Anybody around here'll be happy to help. Well, most of them, anyway. Because we've all been there scratching our heads, too."

He looked hard at me, making sure I was paying attention.

"Just keep in mind," he added with emphasis, *"there's no such thing as a stupid question."*

In the coming years, I kept those words in mind.

I HAD GOTTEN OVER my fear of plumbing by the time we went to work running the pipes in the new house.

Many construction decisions are dictated by common sense, and nowhere is that more true than when it comes to plumbing. For example, the waste pipes must be done first because they fill the wall cavities, while the smaller supply pipes can be fitted around them afterward. Common sense also dictates that the work begin from the bottom, since, as in the structure of a tree, the lower pipes are the biggest, and the rest decrease in size as the extremities of the system are reached.

Our starting point was a hole in the foundation. The foundation plan specified openings in the concrete wall. The foundation contractor had left a sleeve that was to be the exit point for the four-inch waste pipe; a corresponding but smaller opening on another wall allowed entry for the pipe from the well that the well contractor had dug as a water source.

The pipe that fed into the exit hole established the lowest point. The toilets, sinks, showers, and tubs would all be located at higher levels so that the attraction of gravity would draw the water and waste matter to the exit pipe and down to the septic system in the yard outside. In plumbing the system, we had to be sure every pipe was pitched at a rate of at least a quarter inch per horizontal foot. Without that downward slope, the water and solid waste would accumulate and clog the pipes.

Most drain systems today are assembled of plastic pipe, and local building code specified Schedule 40 polyvinyl chloride (PVC) pipe. The plastic pipe is easy to work with, consisting of lengths of pipe linked by fittings that bear self-explanatory names like *elbow, coupling, forty-five-degree Y,* and *ninety-degree T*. The pipe is cut to fit and welded to the fittings using a solvent that chemically bonds the polyvinyl chloride components together.

Referring to the plans, we set about figuring out where the pipes should run. Though the plans specified little more than the location of the fixtures in the house, by identifying where the toilets would be set in each of the bathrooms, we could determine where the soil stacks had to rise in the walls behind.

Starting in the cellar, we cut the pipe to length using a hacksaw, connected the pieces together with couplings, and suspended the lines from the ceiling using pipe hangers. When we reached the first intersection, we installed a T, allowing for a branch line to serve one of the upstairs bathrooms and the kitchen sink. The main line continued on to an inverted T at the base of the main soil stack. One arm of the T pointed toward the house drain; the other, to a cleanout, a dead end with a removable plug that would provide access to the drainpipe should it ever become clogged. The leg of the T pointed directly upward and would serve the first-floor bath, the master bath, and the laundry room.

In order for the two soil stacks to rise vertically through the house, we had to saw large-diameter holes in the walls and floor. We also measured twelve inches in from what would be the finished wall, then cut a hole in the floor large enough for each toilet flange, the fitting that lies flush to the floor and to which a toilet would be attached. A three-inch closet bend—a fat fitting that reaches out like an arm bent at the elbow—was then attached to the stack, between the joists, to connect each toilet flange. After the large-diameter pipes were in place, the smaller drainpipes—two-inch diameter for shower and tub, one-and-a-half-inch for sinks—were coupled into the larger ones at T joints.

Roughing in the plumbing is very far from microsurgery. Only coarse holes are required for the big pipes, and for the smaller ones simple slots are often adequate. But care must be taken not to weaken the structure, so in some instances after a pipe had been put in place, we had to "sister" a length of wood to the channeled joist to reinforce it.

In a matter of days, the waste stacks were in place. The supply system was next.

For that, we would use copper tubing. Copper is both light and corrosion resistant, with joints fixed by heating and soldering them, a process called sweating. Sweating isn't complicated. The tubing is cut to length and the ends are polished with an emery cloth. A thin layer of flux is applied with a brush before you slide on the fitting (perhaps it's a T, an ell, or a coupling). The joint is then ready to heat. A handheld propane torch is the tool of choice, and its hissing flame is then worked back and forth on the fitting until the flux smokes. That signals the joint has reached the temperature at which the solder will melt. A length of soft, thick wire solder is reeled off its spool, and when brought in contact with the fitting, it immediately becomes molten. Thanks to capillary action, the liquid solder is drawn into the joint. The solder hardens in a matter of seconds, though the pipe may be hot to the touch for several minutes. The joint should then be watertight.

Like so much in life, sweating a joint can be done well or badly, and an imperfect joint in a supply line filled with water at fifty pounds per square inch will leak. A slow drip or a veritable shower may result, but any leak, minor or major, must be fixed.

There is no foolproof way to make a tight joint, but plumbing requires careful execution. The surfaces to be soldered must be polished until they shine. The heat must be applied evenly. I found that when the flux began to smoke, better joints resulted if I then counted slowly to three before applying the solder. That extra heating time was enough to ensure the joint stayed hot enough to melt the solder but didn't get so hot that the flux was incinerated and unable to initiate the capillary action.

All the practice at the cottage gave me confidence as a plumber, and a sense of my limits: inevitably, one out of every ten or twenty of my joints would leak. That may sound like a good batting average, but it's not good enough. Fixing a leaky fitting was always a hassle: To test even a single joint, the system had to be fully charged with water. When leaks were found, the only way to fix them was to drain all the water out (the propane torch doesn't generate enough heat to get the pipe *and* the water inside hot enough to melt solder). Then the leaky fitting had to be heated until the solder softened; the joint pulled apart; and the separated pieces allowed to cool before being repolished, refluxed, and heated yet again. So fixing one leaky fitting took as long as sweating a dozen virgin fittings.

The well-drilling crew had run the service line into our cellar and installed a pressure tank that would be the plumbing point of origin for the supply pipes. From there we ran the trunk line, the main cold-water pipe, to the ceiling, then elbowed it to run immediately beneath the joists. After a short horizontal run, we installed a T and ran a tributary down to the water heater. Next we ran a pipe from the outlet on the water heater and began running two parallel lines, with the cold-water trunk line running side by side with the hot-water supply line.

Mike had never worked as a plumber, so I did the cutting, polishing, and sweating on the first fitting while he watched. That was all the instruction he needed, and over the course of the next few fittings, we began to divide the duties. I would measure and call out a length; then Mike would cut and polish the end. I sweated another fitting or two before he volunteered to take the torch.

I handed it over and watched carefully, ready to offer pointers on what he might do differently. Without being told, he directed the flame not at the pipe but at the fitting. He moved the nozzle smoothly back and forth, heating the joint evenly. He applied the solder and, in one motion, wiped the excess off with a rag, a variation of his own devising. He executed the job effortlessly.

"That was perfect, Mike," I told him, trying to keep the surprise out of my voice. He was a little flattered, but he didn't say anything. For my part, I was impressed, since developing a little technique at sweating fittings had taken me quite a bit of practice, and frankly, Mike's first fitting looked neater and tighter than the best of mine. Thereafter, he was in charge of sweating all plumbing fittings. It helps to have a knack for things.

TAKING ADVANTAGE OF MY newfound freedom from my computer, I went to the building-goods supplier where we bought most of the lumber, plywood, and other construction materials. That morning, we needed a couple of specific plumbing parts, and I wanted to fit the bits and pieces together mechanically to make sure they were the right ones. While at the store, I did a little shopping. Unscheduled time at a lumberyard is usually educational—in building things, if you look, you learn.

I found myself in one of the lumber barns, following a clerk who was fetching some fitting or other for me from a storage bin upstairs. We had walked through one area where rows of windows stood, lined up like soldiers. In the next room were cartons of poly, rolls of tar-impregnated paper, and a great deal of planed wood. Much of the wood was pine, mostly number one common, meaning it was nearly clear, with no knots and few blemishes. I admired the lengths of clear wood in the way that one always admires the greener grass next door. The stuff was too expensive for most of my needs, as my budget rarely allowed for the 200 percent leap in price from number two to number one.

The clerk had disappeared through a door on the far wall, and I could hear him as he climbed the stairs to another storeroom above.

My eye was drawn to an impressive pile of tongue-and-groove cedar. Unlike the number one pine, the boards were knotty, but they also had a rich reddish hue. The cedar had a pleasant smell, too, although it wasn't the aromatic subspecies used for cedar chests and closets. I wondered if we could use it to line our closets, and made a mental note to talk it over with Betsy.

Moving on to a nearby loading dock with an overhead door at its mouth, I found pallets piled high with flooring. The cedar sighting had been accidental, but looking at flooring had been one of the purposes of my trip.

Flooring conveys a great deal about a house. Some people have an almost worshipful attitude toward old, wide-board floors. Early American colonists found forests dense with large trees, which they harvested for building materials and firewood. New England, for example, had what must have seemed like endless stands of yellow pine, a durable species that was sawed into beams for uncounted mill buildings and into boards for flooring. Many of those trees were immense, producing boards two feet and more in width. The fruit of that harvest survives in houses that feature the wide-board floors that real estate agents like to point out as evidence of the historic and original character of early houses. Unfortunately, the once-abundant yellow pine was harvested to virtual extinction in the nineteenth century. While pine tongue-and-groove flooring is available, today's products bear only a superficial resemblance to yellow pine. The eastern white pine and the ponderosa pine, from the Far West, are softer woods that dent and scratch easily.

So-called strip flooring is commonplace today. Typically of oak, strip flooring has a face that is two and a quarter inches wide, with tongues and grooves milled into the sides. It's hard, durable, and affordable, but strip flooring has been widely used for only about a century. If the wide pine seemed too soft to be practical, then the narrow oak would be out of historical synchrony with our house.

I continued to wander around, looking at the stacks of pine and oak. One dusty pallet in a corner seemed to be something else altogether. It looked like a hardwood. When the clerk reappeared, I asked him what it was. "That's rock maple," he explained.

The name *rock maple* sounded hard. The width of the boards were also a selling point. Instead of being two and a quarter inches wide, those boards were a full five inches. I left with my miscellaneous plumbing purchases and with two ideas: cedar for the closets, and maybe, just maybe, wide maple flooring.

EVEN BEFORE SHE STARTED walking, Sarah had developed an almost insatiable taste for raspberries. One of my fond recollections is of Sarah almost three years earlier. We had settled her on a blanket at about eight months of age within easy reach of a tall raspberry bush. Betsy was picking berries nearby, and Sarah sat contentedly, plucking and feeding herself berries from the lower branches. Her sun hat protected her from the sun, but nothing could protect her sunsuit from the rich red juice of the berries.

Since then, raspberry pancakes have become a Saturday morning event at our house. Many of the casual rituals of our lives had to be abandoned owing to the demands made on my time by construction, but we managed to keep Saturday morning pancakes on the schedule. One Saturday morning that March, pancakes were on the griddle, and Sarah was consuming her share.

Sarah looked up and asked, "Where does maple syrup come from?"

"Remember those pails on the big trees by the road?" She had noticed them the day before, dozens of gray galvanized pails attached to the trunks of sugar maple trees. She had asked me what they were, and I told her they were Mr. Briggs's and that he was collecting sap. Her attention had wandered before I had been able to explain the pancake connection.

"How about if we go look at where sap comes from?"

The timing could not have been better. Mike was off that morning, and Betsy was taking Elizabeth to her doctor for another lead checkup. Sarah was in my charge, and after finishing a few more pancakes, she happily got into her snowsuit.

The temperature dropped below freezing most nights, but the thermometer some days reached into the fifties. That morning the bright sunlight glinted off the patches of ice where the snowmelt from the day before had refrozen. Dirty piles of snow dotted the ground as we left the house.

At the top of the drive, we turned to walk along Stonewall Road. Sarah's size was a stark contrast to the immense sugar maple trees that lined both sides of the road. We examined a bucket on one of the trees. A spout had been hammered through the bark and into the flesh of the tree. The spout had a little hook on its top that held the wire handle of the pail. On top sat a tepee-shaped piece of galvanized steel that acted as a cover.

We continued walking along the road. When we reached the brow of the hill, a familiar pickup truck was visible. It was parked, its passenger-side wheels on the shoulder of the road. The bed held a dozen old-fashioned milk cans.

I saw a movement in the line of trees and pointed. "Who's that, Sarah?"

She looked and, at first, distinguished no one. A moment later she announced happily, "It's Mr. Briggs!"

He was emptying the contents of the buckets into the milk cans as he did each day during sugaring season. He was concentrating on his work and didn't hear us approaching until we were only a few feet away.

"Hello, Charlie," I said. "Sarah, say hello to Mr. Briggs."

She clung to my leg, saying nothing but smiling shyly.

"Cold night last night," remarked Charlie. "Warm days and cold nights make the sap flow," he added as he emptied a bucket that had been three-quarters full.

"Going to be a good year for syrup?"

At first, he was silent as he rehung the pail. "Maybe. But it's going to be Old Charlie's last year," he added solemnly.

The *Old Charlie* was new. He didn't ordinarily refer to himself either in the third person or as old.

"Finding it hard this year, Charlie?"

His face assumed a pinched and pained look. "My back hurts," he responded. "And I've got gas. Terrible gas. Doctor says he can't do much about it."

I nodded, thinking it was time to change the subject.

After a respectful silence, I asked, "What's the ratio of sap collected to the syrup you get, Charlie?"

"Forty to one, maybe more." He said it gruffly, the self-pitying tone in his voice gone.

"Just this morning, Sarah was asking where syrup came from," I told him.

He explained to us (mostly to me) about his sugarhouse, the big pan he had, and the pine he burned to boil off the water. Sarah didn't follow it all, but Charlie clearly enjoyed the recitation.

Gathering Sarah into my arms so that she could say good-bye to Charlie at his level, I asked, "Charlie, are there rock maple trees around here?" It was a question I had wanted to ask since my lumberyard visit.

He gave me a sidelong glance, then smiled, looking more relaxed than he had at any other moment in our conversation. "You're looking at one right here," he said, patting the big tree with the spout in it. "Sugar maple and rock maple are the same thing." He shook his head and turned to the next tree and the next pail.

Walking away, I felt a little foolish. But the sensation passed quickly as I realized that the choice of materials for our floors was obvious. If we lived amid a veritable forest of rock maple trees, then we should have rock maple floors.

A FEW PHONE CALLS later, we had a fair price from a nearby mill. Four-inch-wide, tongue-and-groove maple flooring would cost $2.60 per square foot.

That prompted a few calculations. Our house had approximately twenty-five hundred square feet. The two and a half baths would have tile floors, and the mudroom would, too. They were the dampest areas, and since water will cause wood to swell and stain, a water-resistant surface such as tile would give better service. Some people like tile on their kitchen floors, too, but Betsy preferred wood, believing it to be easier on her back and feet when she had to stand for long periods.

Including a 5 percent overage for spoilage, waste, and mistakes, we would have to order twenty-four hundred square feet of rock maple. With tax and delivery costs added in, the price would be just under $7,000. Given the roughly comparable cost of oak, the price seemed modest enough—that is, until I looked at our budget. There I had estimated the cost would be some $4,000 dollars less.

That prompted a more detailed look at our financial progress. The foundation and chimney had come in as estimated, but little else had. No one element had grown disproportionately large, but lots of line items had run high. My inexperience had resulted in a few omissions, too. The labor cost had grown, since Mike had proved such a reliable and capable workman, and we had kept him busy and even increased his wages to $15 an hour. He would return to landscaping work in a few weeks, but I had put a note in our Christmas card to Mark asking him to return for the summer. He had just written back agreeing to come for seven or eight weeks, so I knew that in June we would again be working long hours together. Both men were worth every dollar paid them, but the labor costs were almost double the estimate.

We were running over budget by about $20,000, even before the extra $4,000 for the maple flooring. If the same ratio of overrun to budget could be applied to the rest of the job, we would overspend our original budget by about $50,000. My ledger also cited a balloon payment of $20,000 that would come due to Charlie for the property in a few months.

When I compared anticipated expenses to our bank balance, the conclusion was obvious. We didn't have enough cash on hand to complete the house; even with the money due me for my writing project, we would still be short. The realization was painful, and I worried by myself for a day or two before confiding in Betsy. We had both been blithely going about our business—building the house, writing a book, tending to two small children. And the money had been seeping away at an accelerating rate.

"Looks like we'll have to get a loan to finish the house," I told her. That was an option we had talked about earlier but had hoped we would not have to exercise.

She looked concerned but not surprised. "For how much?"

"Well, I'm guessing fifty thousand would be enough. We can certainly get an equity loan on the cottage for that. Or more." We owned the place outright, and comparable houses in the area sold for about $150,000, so the transaction wouldn't be difficult to arrange. "Then, after we move into the new house, we can sell the old one and pay off the loan."

She thought that over, then surprised me by asking, "Does it make sense to put the cottage on the market sooner rather than later?"

I considered her idea, and said, "We can't sell one house until we can move into the new house"—she nodded in agreement—"but we can begin to guess at a completion date."

She was patient, waiting for me to finish. But I couldn't come up with an answer. There were many variables, such as the costs, Mike's departure and Mark's return, and a couple of writing assignments in the offing. There were so many tasks to be completed, among them the siding, the plastering, the kitchen cabinets, and the landscaping. We could probably close on a loan in a month or six weeks. Once on the market, the cottage might sell in a week . . . or a year. We might finish the house in a matter of months . . . but how many?

I shook my head, as if to clear it of competing and conflicting challenges.

"Let's just get the loan," I suggested.

"We can talk schedule later," Betsy said, understanding perhaps better than I that this was not the moment.

NO MATTER HOW STEADILY we worked, the process advanced by fits and starts. We grew accustomed to hearing occasional visitors to the construction site remark, "Wow, you're really making progress!" Just as often, however, the same onlookers on another visit would be strangely quiet, bearing expressions that could be read as, *Has anything changed here or did these guys go on vacation?*

The installation of the wiring and plumbing had taken weeks of work by Mike and me, producing little change in the appearance of the place. But we knew how much work had been done. With the rough plumbing and wiring in place, we could move on to closing up the wall cavities by hanging the plasterboard. That would produce a dramatic transformation.

Until the walls are closed up, the interior of a house is a sea of sticks. The partitions are no more than rows of vertical two-by-fours, bare ribs through which the eye can see from one space to the next. The second-floor partitions in our house divided the twenty-six-by-thirty-eight-foot space into the three bedrooms, two baths, an L-shaped hallway, and multiple closets. Yet without the wallboard skin, the space felt like a boxwood maze denuded of its foliage.

We had a pile of drywall on the floor of what would be the master bedroom. Some forty sheets were neatly stacked just where the boom truck had positioned them. It had arrived before the second-floor walls were raised. The boom's great arm had maneuvered the wallboard off the flatbed and, in the slow motion characteristic of most large machines, set the stack down on the floor. Mark and I had covered it with a sheet of plastic to keep the rain off, and the plasterboard had remained untouched for more than six months. The boom truck had saved us the exhausting labor of moving nearly a ton of drywall sheets into the house and up the stairs.

After painting, hanging drywall is often the next skill the amateur home renovator learns. I was no exception, having done a good deal of it on our cottage. Mike had some experience, too, so we went at it with confidence.

First we chose a long wall with no windows or doors because fewer cuts would be required. When starting any job, the best approach is to begin with the easy bits so that co-workers can learn the sequence of steps and gain confidence. We had no formal discussion of who would do what; the best way to determine roles was simply to do it. Occasionally, I would issue an instruction, but we both knew what had to be done and went about figuring the most efficient approach.

Using a tape measure, one of us would measure the height and width of the space where the sheet was to be positioned. As the first man called out the dimensions, the other would translate them into cut marks on a sheet of the plasterboard. After drawing out the perimeter, we measured for the holes that would be needed on the interior of the sheet for plumbing pipes or electrical-box openings.

Once the necessary lines were drawn, we scored the paper facing with a razor-sharp utility knife. Next, with a man positioned at each end, we brought the piece upright on its long side. For perimeter cuts, the plasterboard was then simply snapped along the scored line, and the backing paper cut to free the main piece from the waste. One man would then hold the sheet steady while the other cut the openings with a narrow-bladed, coarse-toothed wallboard saw, which had a handle like a screwdriver's and a narrow, straight blade. Finally, the cut edges were smoothed with a Surform tool, a special-purpose drywaller's tool that's a hybrid of a plane and a file.

The finesse part was followed by the brawn bit. We would maneuver the piece into position . . . tilting, lifting, turning, twisting. When it was almost in place, I usually mumbled the drywaller's incantation: "When I give this a shove, it should fit like a glove." Sometimes it did. But not always. Every piece that didn't fit had to be twisted, turned, hefted, and tilted out of position and recut. In drywall work, such redos exact a disproportionately high toll on the drywaller, both in time and in wear and tear on the body.

Once the fit was right, the sheet would be fixed in position with drywall screws. These coarse-threaded, one-and-five-eighths-inch fasteners were driven by a screw gun. That's a reengineered electric drill with a motor that runs constantly, so that quicker than you can say "drywall fastener," the screw is driven and holds the drywall firmly in place.

After the walls came the ceiling. We began by fabricating two large Ts out of two-by-fours. The crossbars were about three feet wide, the legs cut so that the total height of the T would be an inch longer than the floor-to-ceiling height. Then we cut individual sheets of drywall to size, lifted them toward the ceiling, and positioned the angled Ts beneath. By setting the Ts to vertical, we could set a sheet of drywall into place and wedge it fast to the ceiling joists above. With the Ts bearing the weight, we would fasten it in place with more drywall screws, without having to hold the drywall over our heads.

The work went well. The spectacle of drywallers at work shares something with dance. We measured our footwork. We circled the pile of drywall, bending and lifting, cutting and carrying. The choreography was spontaneous rather than rehearsed, yet as the hours passed, a practiced sequence developed. It's the same process, the same steps; the repetition is there. The skills improve; the coordination grows smoother, quicker; mistakes become fewer; it's a sort of blue-collar ballet, with Mike and me performing as the mechanicals of the dance world.

Another of the pleasures of drywalling is that the job is self-limiting: when you've covered the walls and ceiling, the task is done. Later will come flooring work, the trimming out of the windows, doors, and baseboards, then plastering, and eventually painting. But when the drywall is up, there's a satisfying sense of completion.

At the end of the first day, Betsy came to inspect our work as she usually did. We had the walls and half the ceiling done. No doubt we wore proud expressions at having accomplished as much as we had.

"So whaddya think?" I asked.

Betsy was almost speechless at the change: our future bedroom had, in a few hours, become a room. As she looked around her, she mumbled, "It's great!" To judge from the look on her face, she meant it.

Trailing a few steps behind her, Sarah appeared. She walked purposefully into the room. At three and a half, she had grown accustomed to visiting the work site and seeing it change, but the look that came across her face betrayed utter confusion. She stood frozen in the doorway. She looked around her, first scanning the room at her eye level, then looking up at the ceiling. I held my tongue, waiting for her unprompted reaction.

After a minute, she looked at me. "Where did all the wood go?" she asked.

What a smart question! I said to myself, the proud father. The framing was indeed disappearing, never to be seen again.

A FEW EVENINGS LATER, Betsy asked me, "When do you think we'll be finished?"

We were both excited by the drywall work. The second floor was progressing so well that Mike and I were about to shift from hanging wallboard to putting down flooring.

Once, the plan had been to build our dream house at a very gradual pace, perhaps over a period of five years. That notion had become outmoded the moment Elizabeth was conceived: we imagined that with two growing girls and two parents working at home our small house would burst. Elizabeth's immobile babyhood had offered us a grace period, but she had already begun walking.

I had worked at establishing a target date for completion, but no easy answers had been reached.

"I don't have a date," I answered. "But I've been thinking a lot about what's possible."

Betsy nodded.

"You know, at the same time I both want to be in there and don't want to move into a construction zone."

"No," Betsy said firmly. "We definitely don't want to do that." We both remembered what it had been like living in our cottage. Images of that were imprinted in our memories: the table saw in the dining room, the living room lined with the shiny aluminum-foil facing of the insulation, and plaster and sawdust on everything.

"We've been making good progress. I've been thinking about the flooring. We should probably plaster the bathrooms first so you can begin the tiling. That's going to take a while. Which reminds me, we need to pick out the tile soon in case some of it has to be ordered."

We shifted the conversation to planning the tiling. Betsy had taught herself the craft of tiling, relying on a how-to book and the goodwill of counter people at tile shops. She had begun with the bathroom at our cottage and had subsequently tiled the second bath we had added and then our kitchen counters. She had an eye for selecting the right border to contrast off-white field tiles. She would bring home a couple of choices from a local tile shop; then we would decide which we both liked (her first choice was generally mine, too). After I had done the arithmetic to determine how many square feet of field tile and linear feet of bullnose edging and decorative trim were required, we would lay the job out together. From there on, she did all the work of cutting the individual tiles, setting them in a bed of thin set mortar, and then grouting them.

Betsy shifted the conversation back. "What were you thinking about the schedule for finishing?"

"Well," I began slowly, "lots of things, I guess. One thing I'm sure of is that we ought to establish a target date. That feels right to me. Not that we're sitting on our hands, but sometimes I think that setting a goal makes you work harder. Or me, anyway. Or maybe it just requires more big-picture thinking and better organization."

Betsy didn't disagree. "I think we could use a deadline, too."

We talked through what had to be done, and within minutes we were imagining ourselves living in our new home by October 15, less than six months later. My sleep that night was disturbed, but I couldn't say whether it was due to excitement at the prospect of living in our new home or to panic at all that had to be done.

An Imagined Arcadia

Own your own view.
—Frank Lloyd Wright

Red Rock has one little mystery. Most of the inhabitants of our obscure settlement don't even know about the cairn. Among those who do, no two seem to be able to agree where it came from.

A network of cart roads and hiking trails crisscrossed the rocky ridge behind our cottage. The couple that owned the property rarely visited but had given me permission to hike on their acreage. Before we began to build our new house, I walked there almost every day.

Many stone walls divide the property, as they do much of rural New England. Photos of Red Rock from the turn of the twentieth century show a landscape denuded of trees, with open fields and pastures. Since then the terrain has become densely forested, and only the long rows of stones remain to mark boundaries between acreage once in tillage. Even the stone walls have settled into disorder, having lost definition, flattened by frost and half-buried in decaying leaves.

Each day my walk was different. Afternoon hikes in the autumn offered the slanting light of the setting sun, illuminating the oranges, reds, and yellows of the occasional maples and oaks along the path. In winter, the turkey trails were obvious, since turkeys drag their tails when they walk, producing a tiny trench in the snow, with claw prints on both sides. One afternoon in early summer, a wild turkey hen appeared in my path. The turkey remained motionless and undetected until I was about thirty feet from her, when she moved toward me all in a rush. I was startled to see her, and my flinch probably cued her rapid movement away from me. Mike later explained that she had been protecting her brood, and when she bustled off, she was moving away from her nest, trying to lure the unwanted visitor with her.

My frequent walks had probably taken me up that hillside more than fifty times when one day the noise of flapping wings caught my attention. Following the sound, I turned from my usual path, passed through a thicket, and emerged into a small clearing around a little pond. I glimpsed a pair of fat-bottomed mallards disappearing between two tall pines on the opposite bank.

As I walked over to the pond, I almost bumped into the cairn. It consisted of hundreds, perhaps thousands, of small, flat stones of native shale and slate. At the base, the cairn was a bit more than two feet across. As it rose to some five feet in height, it diminished in diameter. In geometric terms, it was a hybrid of a pyramid and a cone.

In the days after my accidental discovery, I asked around, but no one seemed to know who'd put the cairn there.

Oddly, Charlie Briggs didn't know. He just shrugged. "I know the little pond you mean," he said. "Though it's really a big puddle, dry most of the year. Never saw a, uh . . . what'd you call it? That pile of stones?"

Since Charlie was unfamiliar with the cairn, I asked the property owners the next time they came to town. "Oh, sure," was the response. "Robbie found that up in the woods."

"Robbie? . . ."

"The landscaper. The guy who cut the trails. Robbie Haldane. You should talk to him."

Talk to him I did. But solve the mystery? Well, not exactly.

Robbie Haldane told me he had heard a story about an Indian princess who had drowned in the pond. Her bereaved beau is said to have built the cairn in her memory. But no sooner had he finished telling me that tale than he added that he had heard another possible explanation, too.

In the nineteenth century, he reported, an itinerant tinker is said to have traveled the cart road adjacent to the little pond on his regular rounds. The story goes that when he stopped at the pond to water his horse, he would gather a few stones and add them to the pile. Over a period of years, the cairn emerged. In that era, when many fewer trees dotted the landscape, the cairn would have been obvious, a marker indicating the spring that fed the pond.

Both stories seemed plausible. Archaeologists long ago established that prehistoric peoples often erected such stone monuments at grave sites, and there's also a tradition of using cairns as landmarks in the British Isles. Both Native Americans and tinkers once frequented the area. Yet for Red Rock's stack of stones, no hard evidence exists to prove or disprove either version.

If the stories were unverifiable, they at least led to an introduction to Robbie Haldane, who immediately charmed us. He was the only stonemason I knew who, except on the hottest of summer days, habitually arrived in a tweed jacket. That was just one indication he was not your average stonemason.

When we met Robbie, his Irish accent was still strong. Later, I was hardly the only one to observe that his brogue grew more dense the longer he lived in America. For Robbie, everything had a story. He told this one to a magazine writer a few years ago. "I did a stone wall a while back. A mouse followed me from one stone pile to the next. I'd finish up with one pile and he'd run to the next. When there weren't any more piles left, he moved to the stone wall, and that's where he lived. I was touched. It meant a lot to me."

Betsy and I knew that we would need help with our landscape; we sensed that Robbie just might be a source for the imaginative solutions we needed.

THE HOUSE DOMINATED ITS site, the central block and two wings hunkering down like a set of closely stacked boxes. With the windows in, the structure actually looked like a house, and as I walked up the drive each day and spied it anew, there was a little burst of pleasure. This was *my house,* the house that *I* was building.

As the snow melted, the view had begun to change. Or maybe it was a mental shift on my part from infatuation to familiarity. With April flowers in bloom in our cottage garden, the job at the top of the hill began to seem not half-done but a long way from finished. The house was surrounded by mud, and instead of that little burst of pleasure, I began to feel twinges of concern. Houses under construction always look as if they are at sea, but our proud home, standing tall and alone, seemed powerless, rudderless.

I called Robbie Haldane for a consultation. In the months since we had first met him, I had learned a good deal more about him. We had spent a few evenings with him and his wife, Teresa, and their children. He played the pennywhistle, producing a sound that seems, at a slow tempo, haunting and hollow; at the breakneck pace of a reel, it's happy and voluble. He liked nothing better than to sit about and talk ("good crack" was his chosen Irishism for an enjoyable conversation). He was even known to break into song—especially during an evening of "drink taken"—in a voice clear and expressive.

Born in Belfast to Protestant parents, he had married Teresa, who was Catholic. Teresa's native county Donegal, located on the northwest coast of the Republic of Ireland, became their home until they emigrated to America. Arriving here in 1988, he got a job laying stone patios, but it didn't take long for him to discover that building a craggy stone wall of granite fieldstone was much more fun than the work of setting a grid of uniform bluestone. Before long, he was saying, "I build stone walls without mortar, the way old farmers built them. I use the stones the way they come from nature." Almost by accident he found himself following in the footsteps of his grandfather, William Haldane, who had worked as a landscape gardener on estates in Ireland.

Robbie also found he could look at a construction site and envision how it could be reshaped. He had the benefit of an old-world imagination, in which the mystical monuments of the Celtic past found their way into the gardens and landscapes of his clients. He went into business for himself and listened to his own personal muse. He built dolmens, in which two or more large, upright stones supported a horizontal slab, like some primitive table. On Robbie's jobs, long, narrow stones are often found upright—standing stones that form crude posts or markers and look as if they've been there since neolithic times.

This was the sort of thinking we needed.

Our house overlooked a small depression. The day Robbie came to look, the view from the house was of a broad but shallow mud puddle where the snowmelt from the winter had accumulated. By summer the standing water would evaporate, but the previous year, that hollow had remained thick with reeds, ferns, and even a couple of wild iris, plants that like wet feet. Their year-round survival suggested the chances were excellent that a deeper hole in that ground would fill and stay filled with water.

Robbie had noticed the vegetation, too.

"Are we thinking the same thing?" I asked.

"A wee loch would look grand there," he responded.

For him, the small pond would be a design element, but for me it would have an added advantage. There were no fire hydrants on Stonewall Road, so having a pond nearby to draw water from might save our house in the event of fire.

In Robbie we had found a valued new collaborator. Betsy and I knew that Robbie's ability to dream in three dimensions would be a great help to us.

MY NEXT CALL WAS to the Gardina Brothers, and this time Leo's brother Felix arrived.

First, he employed a gas-powered pump and a large-diameter fire hose. Towed by Felix's pickup, the pump arrived mounted on its own trailer. Felix unhitched the machine and unreeled perhaps twenty feet of flattened hose as an intake to link the oversize puddle to the pump. He ran a second and longer hose as a drain to a spot some fifty yards away. When he started the machine, it spluttered, then issued a few loud sucking sounds. After a couple of gurgles, water began gushing out of the end of the second hose. The water soon formed a brook that followed a meandering fall line away from our property, down the hill toward Indian Creek.

With the pump running, Felix drove off to get his dozer. When he returned a few hours later, pulling the machine on its flatbed trailer, the pump had run out of gas. But the pond had been reduced to a mudflat.

Robbie and Felix talked. Gesturing at the site, Robbie described his mental image of the pond as being longer than it was wide. Felix nodded. "But I want a little tail at the far end," said Robbie, pointing. Felix squinted almost imperceptibly. His expression seemed to say, *A what? Did he say a tail?*

Robbie talked some more. Together they established that the pond should be about five feet deep at the middle, and the sides should rise steeply to the banks. In theory, that would minimize the growth of grasses and reeds at the perimeter. As Robbie described his mental image, we walked into the empty bowl of mud. Robbie certainly had a clear picture in his mind, but I wasn't so sure that he conveyed exactly what he wanted to Felix.

Unfazed, Felix mounted his enormous dozer. He began by digging a central trench through the mire. The mud rolled before his blade, forming a great soggy cylinder that made slurping sounds as the machine pushed it clear. After two or three passes, he would push the accumulated mud up the gentle slope to the front yard of the house. That was the second part of the plan that Robbie was devising.

When we purchased the property, Betsy and I had planned to change the topography as little as possible. I had said as much to Charlie, both because I thought it to be true and because he would be reassured to hear it. But the terrain around the house site sloped downward, especially to the east, where the grade ramped directly to the pond site, so the house seemed to be sitting on a tilted tabletop. Another design might have used such a site to advantage, but our symmetrical house, with its broad horizontal base, looked like it was about to slide gracelessly down the incline and into the mud.

The solution was in the hundreds of cubic yards of squishy, dark soil. Neither Robbie nor I had done the calculations, but he believed that the earth excavated in digging the pond could be used to reshape the topography around the knoll. I knew the result I wanted—a level surface on which our balanced house would sit. In Robbie's mind, a plan for accomplishing that was taking shape.

Working outward from the central trench, Felix began shaping the edges. Angling the dozer's ten-foot-wide blade, he tapered the contour of the pond bottom, driving the machine in an elliptical arc. Despite its sheer bulk and weight, the big dozer with Felix at the controls sliced and shaped the earth with astonishing precision. Periodically he stopped the machine, and with Robbie at the transit and me as the runner with sticks, we measured progress. First we determined when Felix reached the proper depth; later, that the bank around the edge of the pond-to-be was at a consistent level.

As he worked toward the near edge of the hole, a surprise offered itself. An old stone wall had once run near the pond site. We noticed a few stones on the surface, though most appeared to have been picked and trucked off. But as the digging progressed, many large stones emerged. Robbie was animated as he issued instructions on relocating them. He seemed to have a plan, but he wasn't yet ready to share it with us. He just shook his head when asked, and replied, "Oh, surely, we'll use that stone." He was content to leave it at that.

We trusted Robbie, but in looking at those stones, I couldn't help thinking of many new houses I'd seen where the dominant element in the yard was a collection of big rocks that looked as out of place as asteroids. Sometimes they line the drive or the public roadway out front. They might be set in circles, pushed aside to define a boundary, or simply left in a pile. Usually the arrangement was no more than a transparent attempt to deal with boulders uncovered during excavation without going to the expense of having them hauled away. Despite the presence of the obligatory lawn, a few smallish bushes, and perhaps a flower bed or two, such haphazard landscape design telegraphs, *Sorry! Budget exhausted.* We wanted a yard that didn't give the passerby the feeling that the landscape was the builder's last priority. We wanted something that looked established; we wanted it to look somehow settled, if not old.

Betsy arrived after the trucks had left and was greeted by a huge pile of stones, one worthy of Paul Bunyan. There was also a pile of dark brown muck that looked like it might have been the work of Babe the Blue Ox. The stones and the mud virtually filled the front yard, which had become a great, uneven affair. It looked like a field after plowing, only with furrows deep enough for Sarah to fall into.

I had watched the progress and seen the contours of the terrain changing over the course of the day. I walked Betsy down to the marsh, telling her of the machine and the stone, and then showed her the gaping hole. It was large enough to bury a small fleet of cars.

Robbie was gone, so it was left to me to explain his vision. "That's the tail," I finished, pointing to the far end of the pond. "See? Down there."

Betsy's face took on the confused look Felix's had.

"A *tail?*" she laughed.

"Yeah, you know, a little narrowing. A runoff, kind of?" I gave up explaining with a shrug.

Together we stood looking at the hole, then headed back up the slope, making our way through a yard that had become a slough broken only by the pile of stones. We were both wondering whether Robbie's plan—whatever it was—would transform the property as we wanted.

ALTHOUGH WE SAW NO further progress on the landscape in April or May, we weren't really worried. Robbie could be relied on to return in his own good time. As we waited for the next step in the landscaping process to get under way, the dominant feeling was less a concern about the lack of immediate progress than an excitement about what was to come. But we didn't really know what to expect, since he wasn't working to a detailed drawing. He kept his plans in his mind and did his best to convey his vision in words. The picture would gain focus only over time, so we would have to wait and see.

Robbie did return occasionally, offering assurances that he hadn't forgotten us. On one of those visits he tipped his hand a bit. He told us he wanted to build us a ha-ha.

Dredging through college memories, I came up with the English landscape garden and its ha-ha, the ditch at the garden's boundary. Sometimes there was a fence at the bottom of the trench. More often, the near bank consisted of a steep stone or brick wall, the far one a grassy slope. From a vantage inside the garden, the ha-ha was invisible, disappearing below grade. This dry moat had an important function: to keep the beasts of the forest out of the garden.

I must admit to feeling a little bit proud of myself at being able to call up the image. But at that moment, the ha-ha seemed like a clever solution from another era. What was the connection to our site?

"So . . ." I began slowly, "you think we should have a trench?"

"Not at all," Robbie replied quickly. He said it in the Irish way, in which the words run together, laughingly, as in *NAWT-a-tall.*

"No, I mean we'll build a stone wall, using your stones, a wall you won't even see from the house."

I absorbed that.

"How tall would it be?" I asked.

"Seven feet, maybe eight."

We would build a tall wall that we weren't supposed to see. That sounded strange.

Robbie explained. "Look there." He gestured to the pile of dirt. "If we put a curving wall along *here*"—he shaped a flattened S curve with his palm—"then we could backfill all that dirt behind it."

I began to see the light. "And we'd have a flat expanse of lawn in front of us?"

He nodded, proud of his solution. "It'd be a bit like a giant's billiard table."

In one bold, old-world stroke, Robbie was resiting the house onto a level plain perfectly suited to its breadth and balance.

The solution was unexpected, but the ha-ha is always a little surprising. Today the word is consistently spelled *ha-ha,* but in eighteenth-century usage, it appeared as *Ah, Ah* and even *Ha-hah.* Upon visiting Stowe, perhaps England's greatest eighteenth-century landscape garden (with its *three-mile-long* ha-ha), Thomas Jefferson remarked, "The inclosure is entirely by ha! ha!" The exclamation points that he added contribute suitable emphasis: whatever its spelling, the term is thought to be derived from the interjection *ha,* an expression of wonder or surprise. No doubt many an unsuspecting visitor has said exactly that upon suddenly finding a great trench in his path.

Robbie felt at home working in the tradition of the English landscape garden. English gardeners in the eighteenth century constructed elaborate landscapes around their country homes that were a hybrid of the "natural" and the "artificial." The ha-ha made that marriage of the wild and civilized possible. It not only kept the manicured inner acres free of wildlife but also gave the observer the illusion that the seemingly untrained terrain beyond the ha-ha was part of an unbroken expanse, incorporating the outer wildness as part of the inner garden. Landscape gardeners like Lancelot "Capability" Brown (he liked to tell his clients that their grounds had unrealized potential, or "capabilities," as he put it) emphasized natural effects. He composed clumps of trees, terraces, ponds, hillocks, and great grassy greens into a painterly landscape.

As spring gave way to summer, Robbie Haldane's attention was focused on other jobs. He would be back to work with us, but as he had explained, nothing could be done until the fill had dried. So we waited while the great wave of mud that sat before our house gave up its moisture to the sun.

WHEN ROBBIE LEFT, MIKE went with him, resuming his preferred summertime work of laying up stone, moving dirt, and performing the other labors of the landscaper. For a few weeks, I worked alone on the house, putting down some of the maple flooring and beginning the finicky one-man work of trimming out the windows and hanging the doors on the second floor.

We were all eagerly awaiting Mark's return. When we got word exactly when he would arrive, we decided to make a small occasion of greeting him at the train station. Betsy got the girls dressed in sunsuits and we piled into our minivan, heading down to pick him up on a beautiful June afternoon.

When Betsy saw him, she gave him a hug, and Elizabeth offered a big smile. Sarah clung tightly to my leg, peering out at Mark but shyly avoiding his gaze.

We drove to a hilltop overlooking the train tracks that had brought Mark the two hours north from New York City. The place, a historic site owned by the state of New York, was a parkland surrounding a house called Olana, the "fortress on high," as the original lady of the house translated the name.

We took Mark there that evening because of the view: the vista from the home that once belonged to the American landscape painter Frederic Edwin Church makes it the perfect place for a picnic. The house itself lies just south of the summit of Long Hill. The peak rises sharply some four hundred feet above the Hudson River, with an unobstructed vista of the Hudson stretching south as far as the eye can see.

We spread our blanket on a terrace of grass immediately in front of Church's sitting room. We brought Mark up to date with the happenings of the past year. He was eager to see progress at the house, and Sarah, her shyness wearing off, told him of the mire that had begun to fill with water. "Daddy says maybe we can swim in it!" she told him excitedly.

Immediately in front of us was Church's own pond. He had thought the view of the Hudson River asymmetrical, so he had a ten-acre pond dug to balance the vista. In the 1870s, though, he didn't have Felix Gardina and his dozer to dig his pond, so he paid laborers a dollar a day to excavate a meadow. They moved thousands of square yards of soil with shovel and wheelbarrow.

Betsy had prepared fresh pesto that afternoon, knowing it was a favorite of Mark's. Sarah ate the long noodles in her usual style, bringing larger mouthfuls to her mouth than it could accommodate, leaving strands hanging out over her jaw. Mark said with a laugh of recognition, "I remember that Medusa look." He seemed as glad to be back as we were to have him.

Being at the top of Church's hill with Mark and the girls made me think about the good fortune of my fatherhood. In contrast, Church's life had a couple of fateful turns.

As the Civil War ended, the tall, lean, and confident Frederic Church (1826–1900) cut a striking figure. He had wealth and a good marriage, and as a young man he had met with unheard-of artistic success. Church had not only one-man shows but one-painting shows. Thousands of people would pay to see what were then called "great paintings," enormous and detailed canvases that would transport viewers to places they would never visit—the North Atlantic in *Icebergs,* South America in *Heart of the Andes,* or the great falls in *Niagara.*

Church and his wife, Isabel, had everything. Then, in March of 1865, a diphtheria epidemic in New York took the lives of both their young children.

As my girls romped on the hill—Elizabeth, now a year and a half, joined three-and-a-half-year-old Sarah in rolling down the gentle slope in front of us—I told Mark some of Church's story, though not the part about the children (a story that takes a happier turn, by the way, as four more children were born to the Churches and all lived to adulthood). I thought he would be interested in the tale of the house and the landscape around us.

The 250-acre property at Olana was the chief work of Church's later life. He worked at designing his house in the same way he did his paintings, making many preliminary sketches. Pencil works gave way to color studies. In the archive at Olana, there are hundreds of drawings and paintings in Church's own hand: elevations, and details, and details of details. When the house was complete, he turned his full attention to the landscape.

The five of us took a short walk around the hilltop because I both wanted to show Mark and needed reassurance that our acres, too, might someday mature from an excavation to a landscape. Church had decorated with gardens and thousands of trees. Mark was impressed when I told him Church had cut seven and a half miles of roads, at every turn shaping a view of the house or landscape. Church had confided to a sculptor friend at the time, "I can make more and better landscapes in this way than by tampering with canvas and paint in the studio."

For me, the story of Olana was poignant. I felt as if I shared something with Church, and not only as a homeowner establishing a personal place. He made Olana to be a safe haven. The worst nightmare of any parent is to lose a child; Frederic and Isabel buried two in a matter of days. He then self-consciously built a "bomb proof" house (his words), insisting from the start that the structure be stone. It protected his treasures: not his paintings (he kept few of those, selling all but one major work) nor his furnishings (they range from tourist junk to an oddball array of objects that appealed to his artist's eye). He was protecting his greatest treasure, his children.

As we packed up our picnic and headed home, I found myself thinking of his work on the landscape, too. He hadn't made extensive illustrations or site plans for his landscape, as he had for his house. Instead, he closely supervised the process on-site. Like Robbie, he had a general plan in his mind, but implementing it involved a series of ad hoc decisions. While our project was much more modest than Church's, his home-building was recognizable to me in many ways.

When we got the jet-lagged Mark back to Red Rock, we gave him a quick tour of the house. He made admiring noises, but he was obviously tired from his journey. Almost as soon as we returned to the cottage, he made his way to the guest room for a good night's sleep. By his body clock, he'd been traveling since about two in the morning. And we all knew he would soon be put to work.

AFTER THE MUD HAD shrunk to a dry, hardened bed, Robbie showed he was true to his word. He arrived accompanied by another man with an excavator. This one was different from the one Leo used to dig our foundation: it had a bucket with a mechanical thumb.

Robbie stood before the machine like an elephant trainer, indicating which stone was to be moved next. His collaborator, machine-operator Bill Warner, would then maneuver the arm of the excavator so that the inverted bucket could be lowered over the chosen boulder. Like a giant hydraulic claw, the bucket and the thumb would be drawn together mechanically in a pincer grip. Robbie would indicate where he wanted the stone, and Bill would swing the arm of the excavator into position, twisting or turning the stone as it was lowered into place. Once it was dropped onto the wall, its final position could be adjusted, using the bucket to pull or nudge the boulder a few inches one way or another.

Some of the stones they moved weighed a ton or more; some were only as large as a medium-sized suitcase. None was regular in shape, so the eye of a sculptor was required to identify which great chunk of granite or shale would fit where. I grew accustomed to the hum of the excavator as Mark and I worked on the house. Over a period of three days, Robbie supervised the construction of our great wall. It was seven feet tall and swept away from the house in an elongated, one-hundred-foot curve. When it was finished, a bulldozer backfilled the void behind the wall, then smoothed the yard in front of the house.

MARK AND I STARTED to work on the exterior of the house. Since he would be with us for less than two months, it made sense to focus on work that was better done by two men. Applying clapboards is one job that goes much more efficiently when done by a team.

Mark had never sided a building before. We also hadn't worked together for almost a year, so a small, five-by-fifteen-foot triangle of siding high up on the far side of the house seemed like a good place to establish our coordination and pacing. It could only be reached by standing on the roof of the wing, but that also meant the area would be almost impossible to see from the ground. If we made mistakes up there, they wouldn't be too obvious.

We hadn't been working very long when Mark called me over.

"What's *that?*" he asked in an outraged tone. He was kneeling, pointing at the joint between the vertical wall and the rising roof.

"You mean the flashing?" I asked, a little surprised that he didn't seem to know what it was.

"No," he replied, his tone impatient. "Look here. Look closer. Isn't that a hole?"

I brought my head nearer the piece of step flashing that he was pointing at. He inserted his finger into a hole that was a little less than an inch in diameter.

It made no sense.

"How the hell did that get there?"

We both stood, arms akimbo, and looked at it.

We would have to pull that piece off and replace it; if we didn't, a large volume of water would flow through the hole when it rained. What was not so obvious was how the hole got there. It was much bigger than a nail hole. A hole that size might have been made with a drill, but I could tell that this one had not been drilled because at its edges there were tattered points of copper. This was a puncture: something powerful had been driven through the copper and the layer of plywood behind it.

Mike was within earshot below us. He had returned with Robbie and was smoothing topsoil with a wide aluminum rake. He was dressed in his usual warm-weather outfit, a T-shirt and a pair of hunter's trousers in a green-and-gray camouflage pattern.

"Hey, Mike, c'mon up here a minute, will you?"

Without a word he moved to the ladder, and a few moments later he was standing with us. Mark did his pointing once again, and Mike knelt to examine the hole.

"Are you thinking what I'm thinking, Mike?"

"I think so, but I can't believe it," he mumbled. We were looking down at the top of his baseball cap as he shook his head.

Mark hadn't yet guessed, and he looked impatient.

"You tell him, Mike. You're the hunter."

Mike stood up and smiled sheepishly. "Somebody's shot the house, Mark."

Mark looked stunned.

We removed the piece of flashing and found a deer slug from a rifle inside the wall cavity. The chunk of lead was as big as the end joint of my thumb. But the mystery remained: How could a hunter mistake a tall gray box in the middle of a large clearing for a deer?

The trajectory of the bullet had been from the southeast. In that direction was the ha-ha and, beyond it, a stand of tall pines with a dense thicket at its feet. Though not visible through the lush green foliage, further in, there were several dozen large trees that had been felled by a big wind.

About a year earlier, a tornado had bounced through Red Rock, ripping up trees as a gardener might weeds. None of us had been in the woods to see the trees fall. We hadn't even discovered the damage until several months after the wind had struck. In the same way, when the errant rifle shot had struck the house, no one had been there to hear the gunfire or the loud thump the bullet must have made when it struck the upstairs wall. But those were accidents of timing.

Like Frederic Church, I was constructing what I hoped would be a safe haven for my wife and children. Yet that bullet and those felled trees reminded me how little control we really have over our world and how near are the dangers, both natural and of our own making.

BY MID-JULY, THE ha-ha and the grading of the lawn were finished, though Robbie wasn't. His landscape had other features unlike what you might encounter in today's typical landscape plan. Some of them Robbie had, in his own way, described to us. Others seemed to emerge as the machines circling him rearranged large volumes of earth.

Robbie was working on the open area that rolled down to the nearly full pond. "We'll put in an old cart road," he said, his finger tracing a curving path from the pond to some imagined vanishing point in the woods. In his picturesque vision, there was nothing incongruous about putting in an "old" cart road.

It wasn't a problem for us either, as Betsy and I nodded, agreeing. We stood on the top of the ha-ha wall, surveying the area beyond.

"Just here will be the abandoned tennis court."

The abandoned tennis court?

"This is the big house," Robbie explained in response to our bemused looks. His gesture embraced the house and the modest tableland in front of it. "And a house like this would have had a tennis court, a grass one, gone to seed."

We found ourselves nodding again, and in the next few hours, the blade of a dozer created wavy contours that did indeed suggest that once a cart road had meandered through. Not far away emerged a rectangular flat that really did resemble a tennis court. One day, when it was covered with grass, the feel would be of an old clay or grass court that had been left unused for generations.

LIKE SOME SORT OF upland dike, the ha-ha held back the soil that was our newly leveled front yard. From almost the first moment, it represented much more than a clever reuse of an old idea.

For one thing, it didn't have the sense of artifice about it that would have been familiar to visitors in the eighteenth century. Then, the pleasure garden was in part an illusion, with the immediate garden and the far horizon imperceptibly melded into one unbroken expanse. But Robbie's modified ha-ha didn't really attempt to fool the visitor, since it was very much in view as you approached the house along the meandering drive. The wall was undeniably *there,* forming the bold shoulders on which the house stands, holding back the mass of earth.

This was no high-art wall, either, since it consisted of very large stones set by a larger machine. It had the look of both great solidity and crudeness. If some stone walls look like tightly set mosaics, this one had the tumbled-together quality of a landslide.

Yet it had subtleties, too. A simple stair of large, flat stones snaked down from the upper level to the so-called abandoned tennis court below. That stair was an invitation to move from the upper lawn, which we planned to make a rich green for games and gardens, mowing and trimming it often. The lower green would be wilder, mown only once or twice a year. We wanted the feel of a meadow, with a mix of hay, clover, wildflowers, and wild plants whose seed blew in on the wind or got dropped by birds in flight.

The meadow itself would be a transition. Beyond was the pond and the untouched woods—untouched by us, that is. Much of the forest had been a pasture a hundred years ago, but there was also a steeply sloping hillside with mature hardwoods that suggested it had never been clear-cut by nineteenth-century farmers. Some of the land was high and dry; some acres were low and swampy. The woods around our clearing varied as a city varies from neighborhood to neighborhood, depending upon the topography, the water, and the interference of men in generations past.

Standing atop the ha-ha wall, we saw a loosely defined set of concentric boundaries, of a landscape left increasingly to the control of nature as the eye moved away from the house. Our landscape was in keeping with our self-conscious approach to the whole process of building a historic house that never was. What could be more appropriate, then, than a tennis court that never existed?

I learned much from Robbie and Church. A good landscape looks natural, thanks to trees, plantings, weeds, and even raspberry bushes. But as the historian Simon Schama observed, "Landscape is the work of the mind." There's a good deal less that is natural about it than we might think. For one thing, our perceptions of it are shaped by our preconceptions; for another, its form is often calculated, reshaped by men and giant machinery.

With the wall done and grass taking hold on the lawn, Betsy and Robbie had a conversation about where her perennial bed should go. They talked about where to put it and agreed that the perfect site would be at the rim of the ha-ha. It could be a kind of giant window box: the blooms in the garden would seem to be suspended, with no immediate backdrop to distract from their beauty. That made sense.

By the time they agreed upon this plan, Robbie's imagination had an unstoppable momentum. I joined the conversation just in time to hear him say, "The garden should look like an old lady lost control of it."

Betsy looked shocked, but Robbie rolled on.

"A few vines, overgrown daylilies—we'll make it look old and wild." He was excited by the prospect, but Betsy was far from agreeing that her garden should look as though its caretaker was old and even infirm.

Betsy had stopped listening, but Robbie kept talking. I nodded, following him as he visited the place he envisioned, enjoying the cadence of his speech and his description of the plant materials. The little imaginative ride he was taking me on was good fun. Then suddenly it dawned on me.

Robbie's rich imagination was responsible for something else, too: he had built the cairn. It was stunning how obvious it was and that I hadn't added it up before.

"Robbie?" I interrupted.

He looked at me.

"How long did it take you to build the cairn?"

For a moment, he looked surprised. Perhaps I spied a glint of acknowledgment in his eyes, but if I saw it at all, it was soon gone.

"Not at all," he said, dismissing the suggestion.

Over a period of several months, I asked him again about building the cairn; I inquired at odd moments, trying to surprise him into an admission. He never acknowledged having done it.

Yet I remained convinced he had laid it up, one stone at a time, building a small monument and imagining its mythology as he did the work. Actually, I'm glad he denied having anything to do with the cairn.

After all, every little hamlet should have a small mystery.

Of Columns and Completion

The distinction between past, present and future
is only an illusion, however persistent.
—Albert Einstein

We were working as fast as we could. Betsy had painted and begun to tile the master bathroom, so she had become a full-fledged member of the crew. Work outside had been interrupted by several days of rain, unusual for July, though we had no difficulty in finding indoor tasks to do. More maple flooring still remained to be laid. The last sheets of plasterboard required cutting and fastening in place. A long list of odds and ends kept us busy.

On the third successive day of rain, the sky showed little promise of clearing as we ate breakfast. I felt the need for a job that we could finish, that would offer at least a fleeting sense of completion.

"Staircase day, Mark," I announced. "Something new and different."

The two of us got in my minivan and drove a couple of miles to a friend's house. He had been kind enough to store stair parts in his barn, and after receiving my phone call, he had removed the padlock on its broad sliding door and left for work. Mark and I went inside and climbed a narrow stair to the storage area in the hayloft.

We had to wade through an accumulation of goods emblematic of twenty years of adulthood and a ten-year marriage. There were boxes of college textbooks, an old floor lamp, a pedestal ashtray, and an armchair whose stuffing was leaking through both arms. At the center of the space was a neat stack of lumber that looked to be destined for a specific home-renovation job. Here and there were miscellaneous bits of board and molding that no doubt represented earlier work. We walked around old suitcases and a couple of steamer trunks.

Without difficulty we found the walnut railing from our staircase. The boldly turned newel post, with its bulbous vase shape, was nearby, too, along with a tall cardboard box. The top of the box had been sliced off, and the tapering necks of several dozen turned walnut balusters stuck out like pencils from a can. On the floor were two bundles of what looked like scrap wood held together by loops of clothesline. I looked around carefully to be sure we had every piece that belonged to us, because each part had its place.

We carried the miscellany down to the vehicle.

"Is this all of it?" Mark asked. He sounded disappointed at what he saw rather than relieved at having part of the job done.

Admittedly, there was a sense of anticlimax about loading the collection of old wooden parts in the car. Some of them were broken; many still had bent and rusty nails protruding from them. Taken together, they didn't look very promising.

"We've probably got it all," I responded. "But we can't be sure until we start putting the puzzle together."

Back at the house, we off-loaded the stuff into the main hallway beside the stairway. I had been mentally reviewing the pieces, beginning to consider the logistics of the reinstallation. First we needed to catalog the pieces. Our guide would be the drawings made when dismantling the stair, the drawings that Mike and I had used when installing the stair carriage. I had left them stapled to a bare stud on the adjacent wall.

Going to the foot of the stairs, I reached automatically to where the plans had hung for months. I had walked past the sheets of paper so many times that they had become invisible to me.

They weren't there.

"Mark, check downstairs and see if those drawings fell down there, will you?"

He gave me a blank look.

"You know the ones—they were stapled right here?"

He shook his head. "I don't remember seeing them."

I cursed under my breath, not at Mark but at the growing feeling in the pit of my stomach that we weren't going to find them. That day, at least, we would have to move on to another job.

THERE'S ROUGH CARPENTRY, THEN there's finish carpentry. The tools of the rough carpenter are expressions of strength and power. My pet framing hammer, for example, weighs twenty-three ounces, almost twice what my finishing hammer does.

Framers talk about "killing a nail," and with a weapon like a framing hammer in hand, you could kill more than that.

The framing we had done the previous fall was rough carpentry. The lumber wasn't precisely cut or planed. The fasteners were crude, consisting mostly of large nails. I reveled in the framing, but to judge from my tired muscles and aching joints, framing a house was almost a contact sport. The work involved lots of climbing and lifting, as well as the balance to tiptoe across a half-built superstructure, nailing all the way. It's work best done by young guys like Mark, although I had been able, at least for a finite number of weeks, to maintain the pretense of youthful flexibility and endurance.

By the time of Mark's return, the emphasis had shifted to finish work. If framing is fun and fast and minor mistakes don't count (they disappear soon enough behind the plaster, flooring, and other finished surfaces), then finish work must be precise. Rather than knockout blows, finish work requires pulled punches. The baseboards, the trim around the windows and doors, and the cornice moldings at the junction of the walls and ceiling will remain visible after the house is completed. Finish work is forever on display.

Mark had no more experience with finish work than he had had with framing, so our finishing had begun on the exterior. The clapboard and trim we applied would be covered with three thick layers of oil-based primer and paint. Many joints would be caulked, too. So while the tolerances for joints were tougher than for framing, they were not so demanding as inside. The siding and trim on the exterior would be a good training ground, and I was eager to finish the outside, feeling a bit like the prison inmate who, after digging an escape tunnel for months, finally glimpses a splash of daylight. That the work would bring us palpably closer to completion was only part of the explanation.

For the professional builder, applying the skin to a house is all in a day's work: there's the pride of good workmanship, but the workaday carpenter really only takes home a paycheck. On the other hand, this is a key moment for the designer. It's a time when an imagined image comes to life. What began as a mind picture and became a paper rendering assumes real substance as sawdust flies and the siding gun goes *ka-thunk*. Overhangs cast deep shadows; overlapping boards cast shallow ones. The window and door openings, which were solitary rectangles in the bland field of the fabric that wrapped the house, become frames within frames. Our house would be transformed from a nicely proportioned set of big boxes to a more complex and satisfying assembly of patterns on a variety of planes.

My desire for that gratification was why we decided to finish the front of the house first. We had begun with the small section over the roof where we had found the bullet hole. That had helped us establish our coordination. But the exciting challenge before us was the main façade.

All but the lowest portions of it were out of reach, and at the peak, the house stood more than twenty-five feet above grade. We could have worked from ladders to apply siding and trim, but that would have required moving the ladders many, many times. The only efficient way to do the siding job was to work on a platform that would enable us to move about freely at one level, applying courses of siding in a band perhaps five feet high, before moving the platform up to do the next section. I priced out the options of purchasing or renting manufactured scaffolding, but the costs were prohibitively high. That meant that we had to build our own.

We doubled up two-by-fours, nailing lengths of eight-footers and sixteen-footers together, the joints offset, to produce twenty-four-foot legs. Perpendicular to the legs we nailed five-foot lengths of two-by-six at what would be the top and the mid-point, forming giant F-shaped supports. The free ends of the two-by-sixes were nailed to the house. Other horizontals were added as needed at the height we wanted the platform, and we laid two-by-ten planks across them, providing a solid surface to stand on. Bracing and blocking were added where necessary to ensure the stability of the structure. With our catwalk complete, the ascent could begin.

After devoting much of a morning to building the scaffolding, we went to work on applying the finished wood surfaces to the exterior of the house. We began by covering up the last visible bones of the place, the exposed outrigger rafters that formed the upside-down V of the front roofline. We nailed facing boards (fascia) across the front and soffits to the underside. Next we constructed a long, shallow box that extended horizontally across the façade, connecting the points of the V. This completed the triangle (the pediment) that crowned the house.

Beneath the pediment, we nailed a wide frieze board, its top edge abutting the bottom of the soffit. That joint was, in turn, hidden beneath a molding.

This brought us to one of those carpentry moments I cherish most. We had a good plan to follow, but for several weeks I had been nursing a nagging feeling that there was something missing. We had a problem to solve.

I explained to Mark that there were three evenly spaced openings in each of two horizontal rows across the front of the main block. The plans called for four flattened columns, or pilasters, that would provide vertical divisions between the openings and also would seem to support the triangle above. The pilasters and pediment would then replicate the appearance of a classical temple. I drew a rough pencil sketch on the frieze board as we stood there, twenty feet above the ground.

"You know that handsome white house we went past yesterday? That has the same treatment." Our home was to have a visual connection with nearby houses like that one, which had been built just after the turn of the nineteenth century. Yet I also wanted to echo buildings much further back in time, as far back as Greek and Roman sacred architecture, which consistently used the temple form. As we climbed down from the scaffold to get a better look, the challenge was to figure out whether we could add rounded Roman arches between the pilasters and pediment.

Mark and I looked and talked a bit before I decided to consult my library. Mark agreed to pick up where we had left off cutting and nailing the maple flooring inside the house, and I went to the cottage and to my bookshelves.

My source wouldn't be a nineteenth-century American plan book, nor a scholarly work on ancient architecture. Like so many other builders looking for elegant architectural solutions in the years since the Renaissance, I would consult a work by Andrea Palladio.

ARCHITECTURE DOESN'T TRAVEL WELL. For obvious reasons, buildings are beyond transmission from one city or continent to another. Symphonies, novels, and paintings are movable experiences, but buildings are site-specific.

Yet the work of some architects has traveled through time very well indeed. The creations of Palladio (1508–80) are a case in point. Although he never left Italy and all of his completed buildings stand (or stood) within a radius of some fifty miles of Venice, there are Palladian structures from Berlin to Boston. Cities as varied as Dublin and Philadelphia simply would not look as they do were it not for the son of a humble millstone maker.

The portability of the printed book is one explanation for how a regional style went worldwide. Palladio rendered his philosophy into words and woodcuts in his masterwork, *I quattro libri dell'architettura (The Four Books of Architecture),* first published in Venice in 1570. It was an immediate success, going through a half-dozen editions within a few years. A melding of original designs and studies of classical Roman buildings, *The Four Books* made its deliberate way around the Western world, appearing in translation in Spain, France, England, and Germany in the seventeenth century, and just about everywhere else in the centuries since. There was a great Palladian revival in England in the eighteenth century, where Palladio's grand designs had widespread appeal. People wanted houses that made a statement, and Palladio's formula for using porticoes and pediments and domes on domestic architecture fit nicely. Trickle-down Palladianism reached America, too, in particular in the years after the Revolution.

Another explanation for its enduring popularity is the practicality of *The Four Books*. Palladio's collection of measured drawings constituted a virtual plan book, together with some text that laid out his philosophy. His explicit intention was to enable builders to replicate, in whole or in part, what he had designed. Many of the plates were of country villas to which the hereditary aristocracy and the upwardly mobile alike could aspire.

Young Palladio, born Andrea di Pietro della Gondola (his father owned and made deliveries in a gondola), had apprenticed as a stonecutter at age thirteen and had been cutting, carving, and laying up stone for many years before he began designing buildings. Talking to fellow artisans, then, was second nature to him. Although he built palaces, he was a working man's architect, a designer who was a craftsman first, tied to the ground-zero problems and realities of a builder.

Andrea might have remained a mason were it not for an aristocratic mentor. Count Gian Giorgio Trissino recognized great promise in the young man and gave the twenty-something Andrea a new name to match the great expectations he had for him. The name probably came from an epic composed by Trissino himself. The poem featured an archangel—and architectural expert—named Palladio.

Trissino took his pupil on the first of the five trips Palladio would take to Rome. There, Palladio studied and actually measured ancient Roman temples and other ruins. On his return home, he designed buildings *all'antica* ("in the antique manner"), demonstrating a deep appreciation for the architecture of antiquity. Yet his buildings were no mere copies—he had a visionary talent for building the new, which he exercised for more than forty years.

Most people think of villas when his name is mentioned, but a church is perhaps Palladio's best-known building. One of the most remarkable architectural sites in the world, San Giorgio Maggiore could exist only in Venice. It is a massive edifice of Istrian marble bleached white, set so cleverly on its low-lying island that despite the building's great bulk it seems to float on the Grand Canal, its doors appearing to open over the waves. San Giorgio Maggiore holds a place in the harbor quite like that which the Statue of Liberty has in New York. It's a defining building for its city. When viewed in the fog from across the Grand Canal, this hallmark building seems suspended in time like a Piranesi fantasy brought to life. Yet Palladio's illustrations of the church in *The Four Books* couldn't be further removed from such a vision. They're almost scientific in their detachment, anatomical rather than aesthetic.

Just like the drawings Mark and I were using, Palladio's renderings were of buildings as if seen from straight on. He drew elevations and floor plans, and while he didn't invent such drawings, he was the first to use floor plans and elevations of the same house on the same book page. That innovation made it easier to visualize a building accurately by interpreting or "reading" such drawings.

Our house was already recognizably Palladian. Its three-part configuration of a main block with matching wings made it an obvious descendant of the work in *The Four Books,* which included many such symmetrical palaces and villas. But what I wanted at that moment from Palladio wasn't a design to copy. The planes and volumes of our house had already been established, and anyway, Palladio's own designs were all built on a much larger scale. We needed something subtler.

I wanted to consult the master about making the most of the front of our house. Palladio was the perfect person to go to. In the text of *The Four Books,* he had chosen to use the Italian word *frontespicio* (frontispiece) rather than an explicitly architectural term to describe the faces of his buildings. Palladio used an everyday word partly because he sought to speak directly to the craftsman who might not have the book learning of the educated gentleman amateur, but also because he placed great emphasis on the face of a building.

His influence is easiest to identify from head on. Find most any American house with a center entrance, a triangle at its top, and vertical boards on its corners, and the message is *Palladio was here*. Such structures probably couldn't exist without him —which means more than half of the houses in this country wouldn't look like they do if he had remained an obscure stonecutter. After all, Palladio made it acceptable for the temple front—previously a staple of only sacred architecture—to be used on a house. He established a tradition that countless designers since have adopted.

He understood the buildings of the past as only a builder could, yet he was much more than a mechanic. He reimagined the past, adapting classical elements for his buildings. It's all there in the plates of *The Four Books* for his readers to see and put to use. The sheer utility of his woodcuts accounts, to a great degree, for the long shadow that Palladio has cast over the centuries. The immediate beauty and simplicity of those drawings drew me to him.

As Mark measured, cut, and nailed flooring up at the work site, I sat on a couch, surrounded by books and notebooks, and looked for a design solution. That afternoon the pressure of the process had seemed to be catching up to me, but as I moved from *The Four Books* to a number of secondary sources about Palladio and Palladianism, my browsing and brainstorming seemed like a small respite, a reward for the work of the past months.

In looking at photographs of his work, I saw villas built of brick coated with stucco, with broad, unadorned exterior surfaces. Some had columns or pilasters, like so many early American homes that had come after. It was like comparing apples to oranges, but as I tried to adapt elements for our façade, I also attempted to pigeonhole his villas alongside other historic buildings and to fit my house, too, into the historical context. In trying to cram them all into a logical flowchart, I returned to *The Four Books*.

Something about the family resemblance in those orthographic snapshots struck me anew, and I recognized the essential connection between Palladio and his heirs in America. A key reason for his enduring popularity was a strange irony: his buildings have been less influential than the pages and plates of *The Four Books*. Not that they can be entirely separated, of course, and yes, *The Four Books* features his buildings, and many of them are wonderful, moving, dramatic achievements. But the fact remains that Palladio's *Four Books* has moved vastly more people.

Surely it makes no sense that examining two-dimensional drawings would have more impact than visiting great buildings. Any structure is radically diminished when it's seen merely as a set of orthographic projections. A fleeting impression of a building's three-dimensional character can be gotten from a series of photographs, but buildings are best understood in person. The structure envelops the visitor, who can experience the way the building throws its shadow, the way it commands its site, or the manner in which its pieces fit together. That's different from looking at woodcuts of a floor plan and an elevation reproduced on the printed page.

Still, I believe *The Four Books* would have had much the same impact it has had over the centuries even if the buildings in it had never existed anywhere but on paper as architectural renderings. *The Four Books* was responsible for the Palladian buildings I knew in nearby towns and in the cities of New York and New England. Palladio's buildings themselves couldn't have been the inspiration, as virtually none of his American admirers ever visited the Veneto. While the family resemblance was there, few or no American Palladian houses shared the scale of Palladio's villas, and the floor plans were unlike those in *The Four Books*. After all, the needs of the American inhabitants were very different from those of Palladio's aristocratic clients in the Venetian Republic.

American Palladianism, then, wasn't three-dimensional Palladianism. It was an act of homage in *two* dimensions, echoing Palladio's drawings more than his buildings. The abstractness of Palladio's style, with his reliance upon plain geometric shapes and classical detailing, translated perfectly from the printed page to streetscapes and landscapes all over the Western world. How liberating it has been to many aspiring designers, in many countries and distinctly different eras, to encounter the perfect proportions of Palladio. Not a few neo-Palladians over the years have also appreciated his notion that architecture was about nothing less than how we should live together, as he believed that good buildings and good cities were ennobling. Include me on that list.

The solution we needed hadn't jumped out of *The Four Books*. On the other hand, I came away with a new understanding of Palladio's masterwork and what it meant to me. It was as much a catalog of ideas as it was a catalogue raisonné of the master's work. He intended it to be used just as many designers had employed it in the past. Just as I had, they had been looking to *The Four Books* for help in establishing a tone, a manner, but then felt free to vary the ideas as needed.

After all, good architecture is a mix of drawing upon what came before and responding in an imaginative way to specific circumstances.

I FELT SLIGHTLY GUILTY on going up to the house to check on Mark's progress. My usual policy was to work alongside him, but that August afternoon I put to use my rarely played trump card—I was the boss, after all—and doing so had allowed me to invest a couple of hours in research and rumination and in recouping my energy.

Sarah had come to visit me in my office, and we walked up to the work site together. I tried to explain to her the work done earlier that day to shape the pediment. Kneeling beside her, I pointed upward. "We were working up there today, way up high. We put those new boards across the top."

For a few seconds, Sarah looked and listened attentively, but then she bounded off, her curiosity aroused by a nearby mud puddle. In a moment she was totally engrossed in testing its wetness with the toe of her sandal.

That activity would occupy her for some time. "I'm going up to see Markie for a minute," I called to her. "I'll be right down."

She nodded, the toes of both sandals now in the water.

Inside, Mark's work was obvious. The flooring upstairs was nearly done. All that remained unfinished was a small area of the landing, and that couldn't be completed until the rest of the staircase had been reassembled. Once that was done, we could summon the sanding crew to finish the floor, a job Betsy and I had decided we would contract out to a professional, both to be sure it was done well and to save precious time.

The joints on the flooring Mark had laid were tight and true. "Good work, Mark, you made real progress. Enough for today, whaddya say?"

He said nothing but looked relieved to hear the working day was done. Flooring involves much squatting, kneeling, and creeping about at floor level. "Dinner in an hour," I told him as I descended the stairs. "See you at the other house." Glancing back over my shoulder, I saw him rubbing his knees and stretching his back.

Sarah was no longer at the puddle where she had been. I called gently, "Sarah, Sarah! Where are you?"

She must have broadened her explorations to other puddles. I wondered for a moment at any new hazards on the work site, but no great wave of panic swept over me. We took almost daily precautions to childproof the site.

When I heard no answer to my call, I wandered around the house, moving toward the front, calling to her again. "Sarah! Time to go home for dinner!"

Still no reply.

Rounding the corner of the house, I called a third time. This time there was an answer.

"I'm up here, Daddy," said a happy voice.

That was the moment of panic. I craned my neck to look up at the scaffolding, but at first, I didn't see her.

Then she spoke again. "I'm up *here,* Daddy."

If you had asked her how old she was, Sarah would have told you she was three and three-quarters years old. She had been coached to say that, but she knew she sounded cute and clever when she gave that answer, so she would offer it willingly. But at not yet four, she had no formal training in climbing ladders. That made her a natural, although at that moment, paternal pride at her accomplishment was not foremost on my mind.

She was well out of reach, still climbing the ladder that led to the platform at the top of the staging. Getting from one rung to the next required that she hook her knee and, using both hands, pull herself awkwardly upward until she could shift her weight to the next rung. Momentarily speechless, I watched her hook . . . pull . . . clench with the knee . . . teeter slightly as she got her balance . . . then smile proudly down at me.

My knees had suddenly gone soft.

Trying not to reveal my concern, I called up to her. "Sarah, stay there, okay?"

She looked down.

"I'm coming up."

She shook her head, so I decided that distraction was the best strategy. "Ladders are really for grown-ups," I babbled, "and even grown-ups shouldn't climb tall ones all alone. I'm going to come up and help you . . ."

So much for good theories: being a child with a mind of her own, she started moving upward again at about the time I reached the foot of the ladder. Wordlessly, I climbed quickly and gently, taking care not to shake the ladder, and reached her in what felt like a hour but was more like ten seconds.

My instinct was to grab her and hold her safely. Instead I surrounded her small body, standing two rungs below, my hands gripping either side of the ladder, my arms and body forming a kind of human cage.

"I want to go *there*!" she said firmly, indicating the scaffolding, which was within reach. We were almost twenty feet in the air.

"No, honey," I said firmly. "That's not safe."

She started to object, but a hand on her shoulder quieted her. "No, it's time to go down. You can climb by yourself. I'll be just behind you. Okay?"

Reluctantly she agreed. We descended slowly, Sarah stretching blindly to find each rung.

Reaching her when I did might have been the luckiest moment in the construction process. How easily it could have been otherwise.

THE NEXT MORNING, IT seemed obvious what we should do with the front of the house.

The topmost ornament on the front was an elliptical window Betsy had bought at an auction. Mike had laboriously restored the antique sash the previous winter, and Mark and I had carefully put it in place in the pediment. The window was the work of a gifted millworker, with its muntins dividing the window into a stylized sunburst. It had come from a home built circa 1800 and looked just right in its new setting, like the focal stone in a ring.

The tops of the French doors we had found at the antiques

fair the previous summer also had flattened curves. Betsy and I liked that geometric echo of the elliptical window and had already decided to use similar arcs on the interior arches over openings that would have no doors, like the one between the stair hall and the kitchen at the rear of the house. Most visitors might not recognize the visual connection, but we liked the sense of continuity.

The decision to use the same curve on the façade seemed right, and effecting it wasn't difficult. Using a length of one-by-twelve-inch board, the widest stock we had on hand, I sketched an elliptical line, using a loop of string and two nails. The ends of the curve had to align with the capitals of the pilasters that would frame the bays, so that measurement determined the width. The arch would be as high as the board would allow, since we wanted the arch to be visible from a hundred feet or more away. After sawing, trimming, shaving, and smoothing to meet the pencil markings, we had a convex pattern. We used it to draw three identical flattened curves onto three more lengths of one-by-twelve board, then cut and nailed them in place.

The pilasters were next, and we had to fabricate those, too. Using a three-quarter-inch half-round bit, we routed eight grooves along lengths of pine that were eleven inches wide and a bit more than one inch thick. These flutes would add dimension and shadow lines, making the almost twenty-foot-tall pilasters resemble the columns they were intended to imitate. We made capitals and bases to fit the tops and bottoms and then put them in place, seemingly in support of the arches above.

In Palladio's mild Mediterranean climate, such an assembly of arches and columns would have been freestanding, set at a distance from the building to form a porch or arcade. In Red Rock, though, where cold days far outnumber warm ones, we didn't want a tall porch because it would shade the interior of the house. That would be more appropriate on *Gone with the Wind* plantation homes in the southern United States. Our climate argued for the maximum amount of the sun's heat and light to penetrate the windows of our south-facing house.

Flattening the arcade to blind arches also meant the bold three-dimensional feel of a classical porch became almost two-dimensional. The pilasters and arches we made stood only inches in relief from the plane of the façade. They weren't functional but were there merely for appearance. Many designers and critics believe this to be false or even dishonest—modernists and minimalists place great faith in the logic of what might be called "truth in structure." For them, traditional ornament has no place, and they want their peers to be struck by the originality of their work. But Betsy and I were more comfortable with arches and cornices and columns than with visible skeletons and exposed mechanical systems. We didn't want stark or strange but instead hoped our visitors would be struck by our home's very familiarity. Perhaps Freud spoke for many of us when he remarked in *The Interpretation of Dreams* that the classical style signifies a longing for beauty.

As I fit the last capital into position, Mark nailed the other two in place. As we were finishing, I recognized a song that I had heard playing on the radio a number of times in recent days and had liked. I gestured to Mark not to fire the nail gun as the song ended so that I could hear the name of the artist.

The singer was Shawn Colvin, someone I hadn't heard of. I reached for my pencil. It wasn't in its designated pocket. "You got a writing utensil handy?" I asked, wanting to jot down the name in order to remember it. Mark produced his flat-bodied carpenter's pencil. In bold letters large enough to be visible from the ground, I scrawled "Shawn Colvin" on the arch we had just fixed in place. On the board just above was the sketch of the house front that I had made for Mark the day before.

Until we painted the house the following spring, both the name and the outlines of the drawing were clearly distinguishable from the ground below. Then creamy white oil paint obscured both. Perhaps a generation or two from now, a team of painters will be scraping our house and will discover the name and drawing preserved beneath more layers of paint. They will probably recognize the Palladian proportions of our house, but I wonder whether the name Shawn Colvin will ring any bells.

CONSTRUCTING OUR HOUSE WAS all-consuming. Ten- or eleven-hour days at the work site were common. Increasingly, the evenings, too, were devoted to house-related work. I tried to maintain some semblance of normal family life, but my mind was frequently occupied with sketching construction solutions on paper or worrying about our finances.

Some evenings were spent preparing orders for materials we would soon need. Betsy and I would talk through steps in the process. I made long lists, then went back and annotated them. My several clipboards were thick with sketches, notes, and order forms.

Hinges were ordered from the Woodbury Blacksmith and Forge Company, a smithy in Connecticut that made hinges the old-fashioned way, with a hammer and anvil. We decided upon brass heating registers and ordered them from a factory in Ayer, Massachusetts, called the Reggio Register Company, which casts and polishes the grilles by hand. The supply of nails bought from Wilho and the American Legion hadn't included all the necessary sizes, so cut nails arrived from Tremont Nail, in Wareham, Massachusetts. Glue, dowels, router bits, saw blades, and even molding arrived with the UPS man, who became a regular visitor to the work site.

We began thinking about the conveniences we wanted in the kitchen. Trash bins would be required to separate compost, burnables, recyclables, and trash bound for the dump. Two corner cabinets would accommodate half-moon–shaped units that pivoted out of the corner in order to use space efficiently. We debated how many drawers to have and what their relative heights should be.

The pressure on Betsy was beginning to build, too. She needed to spend more and more time at work on the house as we proceeded with the finish work, since she was the head painter and tiler. The tiling was well under way, but Betsy kept getting interrupted by the need to go shopping. Bathroom and lighting fixtures, kitchen appliances, and other goods needed to be seen, and not merely in the pages of a catalog.

These demands meant that Elizabeth had to be introduced to day care. She was eighteen months old and had just stopped nursing, but not because Betsy decided to wean her. Elizabeth stopped spontaneously, leaving Betsy surprised and a little saddened. The long process of separation is hard for adults, too. But the change made day care a much simpler proposition. So off they went, Sarah and Elizabeth, three days a week.

Elizabeth was still too young to learn to swim, but Sarah went to swimming lessons that summer. Betsy and I were nervous about not being there, but we talked it through. The lessons were well supervised, and the classes small. Missing those lessons would have to be one of the prices we paid for building the house. We consoled ourselves with the thought that Sarah might be more relaxed without us.

The first day Sarah had a scare. She walked tentatively into the shallow water but somehow stumbled and dunked her head.

She was unhurt but frightened. Betsy was told when she picked her up that afternoon that Sarah had simply lost her balance, and Sarah repeated the story to me that evening.

Betsy and I were worried that she might become scared of the water, so we tried to reassure her that it probably wouldn't happen again. When she went off the next day, we encouraged her to have fun at swimming lessons.

That evening, Betsy appeared with Sarah and Elizabeth just as Mark and I were climbing down from the scaffold, tired from working in the sun.

As they trooped toward us, I heard Betsy instruct Sarah, "Tell Daddy about swimming today."

Sarah looked up at me shyly. "I couldn't find it, Daddy," she said, in a voice that carried a strong note of apology. That was a new emotion for her. In her increasingly complicated world, new emotional states were joining the already-familiar ones, such as *I'm happy, I want it,* and *That scares me.* "I looked for it, but I couldn't find it."

I didn't know what she was talking about. Betsy wore an expression that seemed about equal parts mirth and mystery.

"Couldn't find what, Sarah?"

She flashed another new expression, this one impatience.

"My *balance,* Daddy. You told me I lost it yesterday, so I looked for it."

Fortunately I was looking at Sarah and saw her trusting expression, so I managed to swallow my laughter. "That's okay, honey," I said with a reassuring smile. "You didn't fall in again today, did you?"

"No," she said, still reading my face.

"Well, then, I guess it's okay, isn't it?"

When Betsy and the girls had gone, Mark and I had a good laugh about Sarah's "lost" balance. But both of us, working at the top of that tall staging, were very conscious of how precious keeping one's balance could be.

We were working our way around the house, applying trim and clapboards. One morning just into September, Mark excused himself for a moment—"Off for a pee, I'll be right back"—and stepped into the nearby woods, where he found a suitable tree.

My tape measure was just retracting, with its metallic zipping sound, when Mark's head reappeared at the level of the staging where I was standing. I noticed his grin immediately.

Ordinarily, Mark was not a grinner. He liked a good laugh and had a sharp sense of humor, but his demeanor was typically sober. At that moment, though, he looked as if he were privy to a rich but sly joke.

He carefully stepped off the ladder, steadying himself on the platform before he stood up. He reached into a pouch on his tool belt and proffered several folded sheets of paper. The wrinkles and splotches suggested they had been rained on.

He handed the papers over. The pencil drawings were slightly smudged, but what was pictured was unmistakably a staircase. *The* staircase.

"Recognize those?" Mark asked. "I found them in the woods. Caught in the roots of an upturned tree." He paused.

"You're very lucky I didn't micturate on them," he added dryly.

Looking intently at the papers, I only half heard his remark. Although some of the blue ink had run slightly, the numbers that had been assigned to each of the pieces could still be read. I was thrilled at Mark's discovery. Without the drawings, completing the staircase would have been difficult, requiring days of trial and error. Now I had the missing cryptograph in my hand.

"The wind must have blown them out the back door," Mark remarked, shifting his gaze from the stairwell to the tree roots where he had found them.

I felt very lucky—and suddenly eager for the next opportunity to resume work on the staircase.

A few mornings later, on the last full day that Mark would be with us, he was occupied with readying for his trip. So I set straight to work, organizing the parts, sorting balusters, identifying which bits went where. The drawings made that part, at least, very simple.

Two of the balusters were broken, and a joint in the railing had separated over time, but the ailing bits were shortly glued and clamped together. The top step was badly worn and had split during removal, so with the old one as a pattern, I fashioned a new tread from a matching piece of fir. Although I accomplished little more, something greater had been achieved. Having fiddled and fitted, I knew what needed to be done and how to do it.

Mark joined me at the site that afternoon. He was ready to work, but I took off my tool belt instead. A couple more hours of work didn't seem very important on his last day.

Together we walked around the virtually completed exterior of the house. We admired the trim and siding we had applied together. I thought I saw in Mark a real sense of pride at what we had accomplished.

"It'll look even better with a coat of paint," I told him. "But that'll have to wait."

I was pleased with our progress and told him so. He accepted my compliments without a word, but nodded in thanks and, to my surprise, broke into a big smile.

All of us were sad to see him go. He had become not only a fine workman but a friend to each of us—Elizabeth would soon wonder where Markie was, Sarah would miss his reading and game playing. Betsy and I, in his two summers with us, had seen a transformation. He had come to us as a mature and self-contained nineteen-year-old boy, but as he left, he was nearing twenty-one, returning for his last year of college. His nervous distance from us had closed, and a confidence and warmth had taken its place. We knew that when he returned—and this time, we all were sure he would—it wouldn't be as a laborer. A friend, yes, and an independent man directing his own life. We saw him to his train, and he was gone.

WITHIN AN HOUR OF Mark's departure, I was at work reinstalling the railing on the stairs. This was a job to be savored, a job that was its own reward, so I made a point of investing good hours. During the long days that Mark and I had worked on the siding, we had worked at a steady pace. The labor was regular and measured, in its way very satisfying. In a single day, we would transform a large plane of the house, a whole wall or section. The progress was tangible, visible to any observant eye from one day to the next. Yet after several weeks of siding, there was a sense of being on automatic pilot.

The staircase work was a change that brought with it a fresh energy. No two operations on the staircase were the same. It had originally been installed by hand, so every piece was a little bit different. Each had its proper place, a place that it had settled into over a period of some 125 years of service. Every element had to be restored to its designated spot, and the whole grafted into the house around it.

The quality of its staircase once said a great deal about a house and its status. In early American houses, upstairs access was often a ladder or, at best, a steep and twisting stair attached like a Siamese twin to the center chimney. As houses got grander, so did the stairways, often becoming an announcement of the homeowner's wealth and station. For centuries, stairways were reliable indicators of how seriously a house was to be taken—or how seriously its builder wanted it to be taken. Our staircase was near the middle of that plain-to-fancy spectrum, a modest but serious statement by capable craftsmen.

I was keenly aware of the importance of the stairs, aware that I was performing the carpentry equivalent of a heart transplant. A staircase controls the flow in a house; the stairs determine much about its safety, style, convenience, and aesthetics. In our house, it would also be a key link in the chronological chain we were trying to make between traditional buildings and our new one.

Our staircase had to look like it belonged, so a number of hours were spent carefully shaping and fitting the baseboard and other nearby trim. When I finished the work, I was satisfied that a match had been made. The railing upstairs butted to a wall, just as it had in its original home. We did nothing to the varnish on the walnut elements, since the finish was original and had the patina of age about it. We weren't looking for new, after all; we wanted it to look as if it had been in situ since construction. That was at the essence of the whole job.

In the quotidian way of the working life, Mark's visit had been eventful. The exterior of our house had new shadings and details. At about the time Mark left, I met up with a member of the crew who had helped pour the foundation a year earlier. "Your house looks great," he told me. "I just drove by last week."

Then he asked, "How long've you been livin' there?"

I stuttered and stumbled a bit before managing to tell him that we weren't, at least not quite yet.

But what a nice notion! I thought to myself, the idea pushing from my mind the long list of things yet to be done. Then the image of the inside of our house welled up: It looked like Santa's workshop probably does as Santa flies off into the night, his sled laden with gifts. Tired but gratified, the elves are left surrounded by the evidence of the race to the finish. In our case, though, there was something besides the mess to focus on. There, in the very center of the place, was the staircase, carefully reinstalled and fully functional, a sign of how far we had come.

The Race to the Finish

It was a handsome modern building,
well situated on rising ground.
—Jane Austen

Building a house is a great bundling of conflicting sensations and feelings. Nothing gets done as fast as you would like, and the costs quickly get out of hand. The quality of workmanship could always be improved, yet there's so little time to go back and redo something just to make it that little bit better. Ideally you would, but the momentum of events carries you on.

There were moments when I felt as if I were running in place: so much work, so much time invested, yet so much left to do. In early September, one of those slumps came along, and I had to focus consciously on the little accomplishments, thinking of them as foretastes of the greater gratifications to come. A sustained building program like ours, stretching over a period of more than a year, requires periodic pep talks: "Every day, at least in some way, we're getting closer and closer." I had to say it to Betsy occasionally, and she to me. It became almost a mantra in the days after Mark left.

On the other hand, there were moments when at some essential level I understood I was doing exactly what I wanted to be doing and doing it well. In moving to the interior work, I entered the "zone," the same zone that sportswriters talk about. The psychologist Mihaly Csikszentmihalyi has dubbed them "flow states," those moments when a task becomes so absorbing that time seems somehow suspended and there's a strong sense of satisfaction.

I had visited the zone as a ballplayer years before. My growth spurt had come in secondary school (in September of ninth grade I was five feet ten inches tall; by June of that academic year, six four and still growing). Not surprisingly, my coordination lagged, and years passed before it caught up. Actually, my athletic skills never evolved to the degree that many of my friends' did.

Although basketball was what I liked best about adolescence, my playing peak came later. A couple of summers after high school, a group of guys I had played ball with reunited on a summer-league team. These were the same fellows with whom I had chased girls, drunk beer, and encountered marijuana (we all inhaled, by the way). We had a core of six or seven players. I was the tallest, though not the strongest, fastest, wiliest, or best at anything to do with the game of basketball. But we were a good team.

My statistics that summer were respectable, as I averaged

about ten rebounds and nine or ten points a game against lackluster competition. No one kept track of blocked shots, but I had a couple every game. We all played hard defense, ran the court on the fast break, and got some easy points on layups. A couple of teammates shot the ball very well, but I was able to grab a few errant shots and put the ball back up and in. To anyone watching with no particular rooting interest, the tall guy was no more than a journeyman center who, doing nothing either spectacular or embarrassing, played pretty consistently on both ends of the court.

I had the time of my life. Just throwing an outlet pass to a buddy for a breakaway, I knew this was as good as it gets. Not that there was time to stop and think about it: it just felt so natural.

Twenty years later, I found they were surprisingly the same, building our house and playing basketball. There was a time factor in each: the game clock and our building schedule. There were rules (or plans) to follow; there were referees (or building inspectors). Yet the real commonality was more internal than that.

The pleasure of the game wasn't in winning, any more than the pleasure of our house was in finishing. The joy, the bliss, the thrill was in *doing* it. It was in the ability to look ahead in time, anticipate the next move, and then execute it, sometimes very well, sometimes not. The best moments of all were those when I felt that exquisite sense of *I did it*. Whether they occurred after a basketball play or after the completion of a difficult carpentry task didn't matter. The exhilaration struck all in a rush, a moment of unadulterated pleasure.

Then it passed, as fleeting as a single firework in the summer sky. I moved on and got back in the flow with no more than a pause for recognition.

AT TIMES, AS NEW deadlines approached and pressures built, I also wondered why I had gotten myself into this.

I paid the bills every Saturday. After the raspberry pancakes, the ritual called for me to surround myself with paperwork, calculator, and checkbook and do my accounts. My focus was usually just on money in and money out, and the numbers would be crunched in less than an hour. One Saturday in September, however, I took the time to examine the larger picture.

It wasn't pretty.

Several months earlier we had gotten our equity loan, but we had been spending down our credit line faster than anticipated. Our elaborate landscaping had cost more than twenty thousand dollars, and given my strong opinions about lackadaisical landscaping, it was embarrassing to discover there had been no line item on my original budget for completing the yard. We had spent more than three thousand dollars on tile for the bathrooms, and again, no money had been allocated. Just about all the items that had been estimated seemed to be running over budget.

Checking and rechecking the figures showed that we had just about enough credit to finish the job. Yet that also meant that the fifteenth day of each month would bring with it the obligation to make an ever-larger mortgage payment, and it was already more than a thousand dollars a month. The conclusion was obvious: the time had come to adopt Betsy's notion and put our cottage on the market.

The trick would be in the timing: we couldn't sell the cottage until the new house was habitable. On the other hand, the cottage might not sell right away, and when (or if) it did, at least two months would likely be required between the handshake and the formal transfer of the deed, given the usual inspections, mortgage applications, and other time-consuming steps in the process. We would have a grace period but couldn't know how long it would be.

I explained the situation to Betsy. She agreed with me immediately about putting the house on the market, but to my surprise, she asked, "What'll we do for Christmas?"

Christmas had been a peripatetic event in our married lives. We had spent Christmas Day in different places, moving along with members of my extended family from home to home. One year we were at a niece's house in Vermont, the next at a brother's in Massachusetts. But each of these junkets required long journeys in the car. That was a less-than-perfect way to spend the day that we both regarded as the most important holiday of the year.

"We obviously won't be finished by October," Betsy remarked.

No, I agreed. That old target date was no longer likely. There was too much left to do to complete the job in the next few weeks.

"What about by Christmas?"

I thought it over. Plastering. Cabinets. We would have to get the certificate of occupancy. Bathroom fixtures. Painting, lots of painting. The list of tasks was long but not impossible.

"Yeah," I said with growing confidence. "We could do that."

Betsy nodded. "Then we could put the house on the market now."

She hadn't changed the subject after all. We were still talking about selling the house we were living in, speculating on when we could live in the new one, and I realized we had simultaneously been planning a family Christmas party.

"Let's do it," I said.

THE SCRIBBLED LIST OF the jobs to be done was long.

We had to finish hanging the wallboard and putting down flooring on the first floor (those jobs had been completed upstairs). The trim around the doors and windows, the baseboard, and the cornice moldings, too, all had to be cut and fastened. The plastering could then be done, a major, time-consuming task. The wires and boxes for the electrical system were in the walls, but individual plugs, switches, and light fixtures awaited installation, one wire at a time. The kitchen was another large job, involving cabinetwork, countertops, and appliances.

Betsy was progressing well with the tiling, having finished the master bathroom and having begun the girls' bath. But she was nowhere near ready to start the painting, and we both knew that would take a staggering amount of time. Everything had to be primed, and then two coats of flat paint had to be applied to the walls and ceilings, followed by two more coats of semigloss on all the moldings.

That's what we had left to do, along with a much longer punch list of minor tasks. And all by . . . *Christmas?* That was a bit less than four months away.

We needed assistance, and there simply wasn't time to break in new help. We needed Mike.

"I can work weekends for now," he responded. "And more as landscaping work slows down." He sounded pleased at the prospect, and so was I. He was a reliable and hard worker.

Some of the crew jobs couldn't be undertaken until certain solo jobs were completed. Mike and I could plaster efficiently as a team, but not until the trim was in place. My Monday-to-Friday energies for a few weeks would be devoted to finishing the interior trim, room by room. The plastering could be done on weekends.

Plain and undecorated can be a design statement—like a birthday present wrapped in plain brown butcher's paper or, say, a house without decorative touches. But that definitely wasn't our style. We intended to complement the exterior finish—the elaborate window trim, blind arches, and pilasters on the front of the house—with a range of moldings inside.

Moldings have regular channels or projections and can be of wood, plaster, or, on the exterior, stone or terra-cotta. As well as being decorative, moldings are also transitions from one surface or material to another: A baseboard molding appears to join the floor to the wall; the cornice, to connect the wall to the ceiling. Casing or architrave moldings finish off window and door openings. Some moldings perform practical functions, like the chair rail around the perimeter of a room at chair-back height, which protects the plaster. For that reason, among others, we decided to have a chair rail in our dining room.

My fondness for moldings has less to do with their utility than with their appearance. Moldings add definition to spaces, in the way a mat frames a photograph. Under the softer illumination of candlelight, the shadow lines and the three-dimensionality of traditional moldings are revealed. Historically correct moldings also convey much about the chronology of a house: some molding schemes proclaim *Georgian!* Others bespeak the Federal, Greek Revival, or Victorian styles.

In the old days before the advent of the power planer, every piece of wood had to be planed by hand. That's no longer true, as our boards arrived from the lumberyard already planed. I purchased stock moldings where possible, too, but not everything arrived ready-made. The table saw was needed to resaw some boards to half their stock thickness. A router rounded one edge of the baseboards to a half-round profile, or bead. A couple of other molding profiles were milled on site with the router.

Mike's weekends-only availability would give me the freedom to concentrate on the time-consuming trimming-out of the windows and doors. Much of that work was trial-and-error, cut-and-recut work better done by one man than by two.

The real satisfaction wasn't in making the moldings; it was in cutting and installing the various parts. Consider a door, for example. At the beginning of the process, there's a rough opening in the frame and an old door leaning against a nearby wall; at the conclusion, the door swings open and shut, alternately clearing and filling the opening. Around the doorway, a molded frame completes the conjunction of door and wall.

The installation begins with the two side jambs and the top jamb. Each is sawed to the correct length. A three-quarter-inch groove is then cut into each side jamb in order to allow the top jamb to be slid, glued, and nailed in place.

Next, this assembly is nailed into the rough opening and becomes a part of the structure of the house. The rough opening isn't precisely the right size, so shims and coarser blocks of wood are used to set the frame correctly, with the sides plumb, the front edge flush with the plasterboard wall, and the top level.

Next are the stops, the boards that are positioned inside the frame, leaving a space the thickness of the door on the swing side of the opening but extending out on the other side to set flush with the plaster. That means each stop must be ripped along its length to the proper width, then cut to the proper length. The top ends of the side jamb stops (and both ends of the top stop) are cut, or mitered, at a forty-five-degree angle to fit neatly inside the frame.

Next, the casings are applied on either side. We had decided upon a complicated scheme, involving plinth blocks at the bottom, impost blocks at the top, and one-by-six side and top architraves. After those elements were in place, we would apply beads on the inside of the jamb and two sets of trim moldings on the architrave. Altogether, that represented thirty-eight parts per door and at least fifty cuts. The first door I did took me several hours, but as I found my pace, the time per door dropped to about an hour.

Tolerances were tight. Every piece had to be cut quite precisely. A piece that is a hair short or a hair long just won't work. There's even a distinction that some carpenters use—quite seriously—to specify how much to shorten a piece that is just a little long.

"Trim that down a blond one, will you?" means a tiny bit, perhaps a sixty-fourth of an inch; "It's a red one long," means it's probably more like a thirty-second of an inch long, since red hair tends to be coarser than blond hair. It's silly, perhaps, but useful.

Even skilled finish carpenters make lots of mistakes. The ability to cover up those mistakes with a minimum of extra effort and materials is at the heart of good carpentry. All carpentry requires common sense, but finish work requires the proverbial slyness of a fox when it comes to correcting mistakes. When a piece is a little long, that's no big deal. A mumbled curse, a second measurement, another cut, and the problem is forgotten. But not all mistakes are that easy to rectify.

Sometimes the mistake is in how several pieces fit together. Let's say an assembly of window trim is all ready to glue and nail in place, and *Damn it,* the realization strikes that something is wrong. Perhaps the pieces don't align. Or you've just run out of a certain molding. At such awkward moments, the genius of carpentry must surface and help solve the problem without remaking the pieces that have already been cut. That would waste time and materials, the carpenter's most precious commodities.

One trick of the trade was borrowed from the magician: The

eye is easily fooled. Strict symmetry isn't necessary; not every piece or every joint has to be identical. If you look carefully at a clapboard house from a distance, for example, the clapboards usually appear to have the same show, meaning the distances from the horizontal butt of one clapboard to the ones above and below are the same. The show looks consistent—but it usually isn't.

Look again. The carpenter relies upon the eye to be distracted by patterns of lines and will vary the show on clapboards in order to align the butts neatly with the bottoms and tops of any openings in the wall. The show gradually changes on our house from a bit more than three inches to four inches, but the brains of most observers register the clapboards as identical.

I enjoyed the finish work, only partly because it's a blend of problem solving and optical illusion. Another of its pleasures was the satisfying mix of low tech and high tech tools. Two tools typify the finish work: the block plane and the table saw. They're at the opposite end of the spectrum. The table saw couldn't have existed until precision machinery was made possible by rapid advances in technology in the nineteenth century. The block plane has been in use in more or less the same form since its invention by the Romans at the beginning of the Christian era. The table saw makes cutting stock to exact measurements very easy; the plane makes erasing any sign of the saw's work (rough saw marks) equally simple.

For much of September, I worked on the trim. The baseboards, cornice moldings, and casings around the windows and doors went up, room by room. They had to be finished first because they defined the perimeter of the areas we would plaster. When the trim in a room was completed, Mike and I could come in and cover the exposed wall with plaster, like children working in a coloring book, neatly filling in the open areas.

Since the late eighteenth century, plaster has been the most common interior wall surface. That made it the obvious choice for our new-old house. Yet not all plaster walls are created equal.

Until the twentieth century, plaster walls consisted of three coats of plaster applied to a series of thin wooden strips called lath. Lath was typically made of softwood, cut to an inch or two wide and a quarter of an inch thick. The wooden lath was nailed horizontally to the vertical framing members, or studs, leaving gaps between the strips. The first layer of plaster, called the scratch coat, was rich in lime, typically mixed with cow hair to reduce the tendency of the plaster to crack owing to the inevitable flex and movement in the structure over time. As it was troweled on, the thick plaster would ooze though the gaps between the strips of lath to form keys, which, upon hardening, would wedge the plaster surface mechanically in position. A smoother layer called the brown coat came next, named for its color, the result of the sand in the mix. The creamy white finish coat contained neither sand nor hair: it was still smoother and could be tooled to a mirrorlike finish. Typically these three coats were applied on consecutive days.

That was the old way. In today's hurry-up world there is, of course, the shortcut of wallboard. Plasterboard, gypsum board, and Sheetrock are only some of the various designations for the same material. Still another, drywall, may be the most descriptive name.

Drywall is a factory-made sandwich of gypsum with layers of paper on the exterior surfaces. As distinct from the "wet plaster" of the traditional three-coat plaster-and-lath wall system, drywall arrives on the work site dry, typically in four-by-eight-foot sheets. Its advantages are obvious: Most any room can be "'rocked" in a day or less. The dimples on the surface at the fastening points, where the drywall screws or nails have been recessed, as well as the joints between the sheets, are then covered with the premixed plaster product familiarly known as joint compound. Because it's economical and both quicker and easier to install than traditional plaster, drywall now accounts for virtually all plaster walls in new construction. In fact, when a drywall contractor is doing the work, a crew can complete a wallboarding job at remarkable speed: with one team hanging the drywall and another applying fast-drying compound, a modest drywall job can easily be done in a day or two.

But the end result of even the best drywall job isn't the same as traditional plaster. Wallboard provides little soundproofing. A sharp knock on a gypsum-board wall produces a hollow sound that carries through the wall. The same rap on a plaster wall produces a dull thud that seems to die away in the masonry mass of the plaster surface. Three-coat plaster is virtually soundproof. It's probably significant that you can put your fist through drywall. Through three-coat plaster? Maybe Mad Mike Tyson could do it, but I wouldn't want to try. It isn't quite true to the cliché *hard as a rock,* but it's close.

We had hung drywall for budgetary reasons: three-coat plaster is about three times as expensive, too. As usual, though, we had had an intellectual wrangle with ourselves about the wall surfaces before coming up with an affordable means of having much of the economy of drywall but some of the advantages of traditional lath and plaster.

Having done some experimenting with plastering at our cottage, we knew that a skim coat of real plaster applied to a drywall surface added considerably to the soundproofing of plain plasterboard. When its entire surface had a troweled coat of plaster, a wall looked better, too. Not everyone agrees, but I think drywall, when hung and compounded well, looks too perfect. It looks somehow unnatural. When done badly, however, it's worse. The smoothness of the board surfaces contrasts sharply with the joints between. The sloppy look of a bad compounding job is unmistakable.

In comparison, a plastered wall, even if it's no more than a skim coat atop wallboard, looks like handwork rather than machine work. The best plaster job has variations in it, waves that raking light makes apparent but that the naked eye reads as the work of the human hand. I wanted our house to communicate the work and energy that went into it; the plastered walls would have the handmade quality of a house from a time when almost nothing came prefabricated from a factory. There was another reason to opt for a layer of plaster: I like plastering. Physical work in general agrees with me, but the fun quotient varies. Digging post holes, for example, or doing roof work requires little skill and wears out the back and spirit, and at the end of the day, what have you got besides sore muscles?

Plastering is different.

Not that it's easy. Plastering, too, is hard work. Big bags of plaster and buckets of water must be hauled to the area that is to be plastered. Once mixed with water, the plaster is heavy, and spreading it high and low, and then smoothing it, leaves the plasterer bone-tired after a day's work. It's real work—yet plastering is a pure joy.

The plaster we use today is usually of gypsum. It's made from the mineral calcium sulfate and is sometimes referred to as French putty or plaster of paris, since its original source was the deposit of gypsum beneath Montmartre. Produced by heating rock gypsum as found in nature to expel the water content, plaster is a fine white powder. When moistened, it hardens, or sets. Technically, this process isn't a matter of drying but a chemical reaction that releases heat over a period of a few minutes.

What makes it interesting is that plaster hardens quickly. At first, a fresh batch slops about on the hawk like a watery cake mix. Dollops will slobber off the trowel, landing with a loud *plop* on the paper-covered floor. In a matter of minutes, however, the plaster begins to stiffen and the race is on. You've got fifteen minutes, tops, to apply and smooth it before the plaster becomes too hard to work.

There are two skills involved in plastering: one is to apply a coat of more or less uniform thickness; the second is to trowel that coat smooth. The plasterer selects an area, typically a four-foot square bounded by a door or window on one side and by the ceiling or baseboard. The plasterer then lays on the plaster, using the trowel in an upward motion. The plaster isn't just slopped on; the idea is to cover the whole area in as few strokes as possible, with a bit more than a one-eighth-inch layer. This goes quickly—only a minute or two should be required to slather the plaster on.

The second process involves revisiting the areas already covered to compress the plaster in long, smooth strokes. Each stroke should slightly overlap the previous one until the whole area is smoother. You do it again, sweeping and smoothing. A good plasterer then moves on to an adjacent area, puts on more plaster, and smooths it in the same way. When a larger area of wall is thus covered with plaster, the laborer goes back and smooths the entire section again.

There are tricks to the trade, of course—lots of them. Just as in painting, you have to keep a wet edge: if you let the border of your plastered area get dry, then a joint will be apparent later. A comfortable rhythm is established only with practice, though it seems to gain speed of its own accord. It can also be enhanced by music.

Mike and I found that hip-hop suited the work of plastering.

Betsy said that watching us plaster to such music was like watching an old silent movie. We became two Keystone Kops in search of a wall to be plastered.

My affection for plastering wasn't born of a sense of mastery. I can lay it on quickly and efficiently—but again, that's only half the task. My smoothing was adequate at best, but to my great good fortune, Mike did it surpassingly well. We evolved a system where I laid the plaster on quickly and Mike did the smoothing. He did a painstaking job, producing a smooth finish.

One weekday when he wasn't around, I decided to finish the ceiling in the closet of the master bedroom. It was the last remaining unplastered surface in the bedroom area, and eager to get the painting under way, I plastered it alone, troweling the plaster on confidently, applying the preliminary layer. It went on easily and looked pretty good for a first pass.

Smoothing came next. It was a rectangular surface, about five feet by seven feet, so there didn't seem to be much risk of its suddenly hardening and becoming unworkable. Yet the more I worked the surface, the more uneven it seemed to get. A few minutes later, as the plaster hardened, every trowel stroke seemed to add another scratch or gouge.

I sponged the area thoroughly, as Mike would have done, and tried to fill the imperfections using the fine slurry of plaster the surface water produced. A painstaking hour brought little improvement. Looking up at my handiwork, I had a feeling of both embarrassment and frustration. My debt to Mike's skilled smoothing strokes was more obvious than ever.

I decided to leave my trowel-scarred bedroom ceiling just as it was. Many years ago I asked a Jewish friend why members of his faith wore yarmulkes. He patiently explained that wearing a yarmulke was an act of submission before Jehovah. Plastering isn't a religious act, of course, but my ceiling is, to this day, a daily reminder of my limitations. While not an act of submission, it was definitely an admission that I don't have all the skills. And looking at it is very humbling indeed.

By the end of September, six months had elapsed since the nurse from the board of health had visited Elizabeth. I happened to come home for a midmorning coffee break to find Elizabeth in midperformance.

She was supposed to be piling small blocks on top of one another. Apparently some psychologist had determined that a child of her age was "developmentally on task" if she could stack seven. Elizabeth had stacked twelve and showed no sign of stopping. She was always keen to please, and she had discerned the delight her stacking gave the nurse. As she deftly set number thirteen on top, she smiled shyly.

The nurse grinned back. All pretense of professional distance had been abandoned. She told Betsy later, "This was such a treat! Most of the kids I visit have real problems. But Elizabeth?" She laughed lightly, and we were relieved that Elizabeth's brief exposure to lead had done no permanent harm.

I decided that if Elizabeth's development hadn't been interrupted by the building process, then perhaps Sarah's might even have been aided.

At day care one day, Sarah was playing with blocks. Miss Diane, the proprietor of Long Barn Daycare, had often reported block play before, but on this particular day Sarah had a larger vision. Her project seemed to be a great, sprawling edifice.

Diane asked her what she was building. "Are you making a house, like Daddy?"

Sarah didn't even look up, Diane reported. Concentrating on her work, she responded, her jaw set, "No, I'm building a city."

In October, we signed a contract for the sale of our cottage. Several dozen potential buyers had come to look over a period of several weekends, but initially our real estate agent had fielded no offers. After a bit more than a month, a couple returned for a second visit, and a few days later, they came again, this time bringing a friend. The following week they made an offer that was in the range of our asking price, and a deal was struck.

The agreed-upon closing date was soon set for December 22, officially establishing the date by which we would have to vacate —and, obviously, move into our new house. Betsy was pleased that we would be able to celebrate Christmas on our turf, and we sent word to our families that we'd host the Christmas Day party. It was a relief that the proceeds from the sale of the cottage would soon solve our money concerns, but the fixed deadline added to the pressure-cooker atmosphere at the work site. We had so much left to do.

The plastering upstairs was finished. The flooring contractor had come and gone, first sanding, then applying three coats of water-based polyurethane to the second-story floor. The painting had begun in earnest, and it hadn't taken us long to realize that Betsy would never be able to do it alone. Since Mike and I were still occupied doing a hundred other jobs, we needed more help.

One morning, Mike mentioned that his wife, Tammy, had lost her waitressing job the previous day. We didn't know her well: I had met her only once, briefly, but Mike said she had done some painting and needed work. Two needs, ours and hers, added up to one hiring decision.

She was a sturdy young woman, with blond hair and a quiet manner. She arrived with paint-spattered clothes, but it didn't take us long to realize she wasn't a trained professional. Although I'm far from a neat painter, the speed with which Tammy's hands got covered with paint each day was remarkable. But she proved to be a willing and reliable worker, though to this day there are a few spots where drops of paint imprinted with Tammy's sneaker tread are still visible.

With Tammy on the payroll, the pace of the painting more than doubled, but the work that remained was still daunting. I really didn't know whether we could meet our deadline or not. But by the end of October, Mike was able to join the crew full-time.

NOVEMBER BROUGHT HUNTING SEASON. In previous years, it hadn't meant much aside from the distant echo of early morning shots in the woods and the sight of deer carcasses stretched across car roofs and trunks. Since I couldn't quite get out of my mind the unknown hunter who had shot our house, the new season seemed different. The piece of flashing with the hole in it was tacked up on a first-floor wall, a bizarre conversation piece.

As hunting season approached, Mike's excitement grew by the day. We talked as we worked, our conversation frequently interrupted by earsplitting saw cuts and hammer bangs. Many of our exchanges were work related. But we talked of other things, too.

Mike told me that he had started hunting at thirteen and that his instructor had been his mother. She had raised him largely on her own, since his father had died three months before Mike's birth. Mike's father had cut himself while cleaning a pet racoon's cage. Initially left untreated, the infection produced a high fever, and by the time he sought help at the hospital, the infection was beyond treatment.

A stepfather had not come until much later, so his mother, Barbara, had taken Mike into the woods. She herself had been taught by her father. She was the youngest of a houseful of girls, and he, despairing of ever producing a boy child, took to calling her Bob and taught her to hunt. There's a cherished family photo of her with her bow in hand, standing next to a doe that's hanging from a tree. In those days, more than forty years ago, deer were rare, and the bows much less sophisticated than those sold today. That she bagged a doe with a bow suggested what an adroit hunter she was.

The mere mention of the hunt made Mike more talkative than usual. You could hear the energy in his voice when he talked of putting out the deer stand, a ladder-and-seat arrangement that would enable him to sit comfortably in the cover of the lower branches of a tree, waiting in the early morning for a deer to happen by.

Mike liked to work—but he loved to hunt. When the season began, he asked if he could come to work later than usual. On the first days out, he saw plenty of female deer but didn't shoot one, even though he had a state-issued doe permit. When I asked him why, the shrug of his shoulders conveyed that, according to some unwritten and probably inexpressible hunter's code, to shoot a doe before bagging a buck was akin to cheating. But Mike continued hunting, confident he would get his annual buck.

About two weeks into the season, he asked me if he might hunt on our property, in particular in the vicinity of the fallen trees just a hundred yards or so beyond the ha-ha. I talked his request over with Betsy. Knowing Mike's careful ways, we agreed to trust his good sense, confident he would take great care. In a strange way, we were looking for reassurance that having a hunter on the property didn't pose a risk to us.

Why Mike wanted to hunt there was no mystery. I had spotted a buck nearby several times in recent months. Once, it had emerged from the swampy area onto Stonewall Road just as we were turning into our driveway. On hearing these reports, Mike had nodded knowingly. "They like tree falls—lots of cover for them during the day. The deer was probably sleeping in there," he explained.

One morning, Mike arrived at the house, parked his truck, and walked into the woods. I wasn't there to watch him; the time was a bit after five o'clock, well before dawn. When I got to the house about three hours later, I called his name, but there was no reply.

Perhaps an hour later a truck pulled into the driveway. It was Charlie Briggs's charcoal gray, half-ton pickup. I went out to say hello, but before Charlie could lever his old bones from his seat, the passenger-side door swung open and Mike got out. A proud smile rippled across his face.

"I thought I had myself a deer," Charlie said, walking gingerly toward me. "Just driving along the road, that deer almost run into me." He jerked his thumb toward the back of his truck. "Then it just collapsed right there in the middle of the road."

A glance into the bed of the pickup confirmed the presence of a dead buck, its antlers big enough to hang a small collection of bowler hats.

"I was wondering to myself, *Can I get him up in the truck alone?*" Charlie continued. "Then Mike here came outta the woods." He cocked his head at Mike and smiled wistfully. It was a look that showed both a sadness that his hunting days were behind him and a happy thrill at the remembrance of them.

"I got him in the lungs," Mike filled in. "Just missed the heart. But he couldn't run far. He must've died just when Charlie saw him."

Helping Mike shift the deer into his truck, I noticed his hands were stained with blood. Seeing my glance, he explained that he had gutted the deer on the roadside. I was reminded of the old phrase *caught red-handed,* a reference to poachers who are identified by the blood on their hands.

When we finished, I turned to Charlie. "Will you come in and have at look at our progress?"

Glancing at the makeshift steps that led into the house, he declined. I made a mental note to add to our punch list the task of fabricating something safer for Charlie and others who needed solid footing and a good railing. With an almost ceremonial handshake of congratulations for Mike, Charlie left, his truck seeming to pick its way tentatively out of the driveway.

And Mike and I went to work on the cabinetry.

OUR BUDGET DIDN'T ALLOW for buying prefabricated cabinets, so making them fell to us.

I had performed the same task before, in renovating our cottage. I had paged through a couple of books about cabinetmaking and looked at a few sets of cabinets and had then gotten on with it. First, I cut sheets of plywood into sides, bottoms, and back pieces, producing topless and faceless boxes, which I then positioned on top of a base. A face frame was applied to unify the separate cabinets and accommodate the doors. The countertop came last.

The job was a bit more complicated than it sounds, and frankly, I could never make a living as a cabinetmaker. The necessary desire for perfection isn't part of my makeup. But doing cabinetwork on the cottage had taught me some important lessons. One of them was how essential drawings are.

Having no more than a handful of plans was rarely a disadvantage in building the house. What wasn't on paper either was in my head or could be resolved with an on-site decision. But the pleasure of devising ad hoc solutions in framing and trimming was transmogrified into absolute frustration in cabinetmaking. Instead of the thrill of devising a good solution, usually there was the realization that we would have to start again or do something else. As the work on the kitchen approached, I had to go back to my computer and do some drafting.

Cabinetmaking is one area where there's no such thing as too many drawings. The discipline of laying lines and dimensions on paper puts questions to the draftsman that if left unasked will lead to false starts, wasted materials, and painful compromises. So many disparate pieces come together in the small space of a single cabinet that mistakes are inevitable. I learned this the hard way.

For our cottage, Betsy had bought an antique marble sink for the master bathroom. It required that a base cabinet be shoehorned into a corner space. The fit would be tight, but the result would be the built-in look we wanted. With a few measurements and a sketch to go by, I had cut out the parts and carefully assembled them, making the cabinet box and a face frame to fit an old door we had bought at an antique shop. Even before painting, the cabinet looked great, and with growing pride and excitement, I maneuvered it into position.

It didn't fit.

After a stunned moment of sheer wonder at my stupidity, a couple of choice expletives, and several deep breaths to bring my anger under control, I remeasured the space and the cabinet. The box was precisely 1 inch too wide: the tape measure had read 36⅞ inches; my eye had registered the fraction correctly, but I had written down the whole number closest to it. So the cabinet was exactly 37⅞ inches wide. That extra inch meant the cabinet had to be remade.

I wasn't planning to repeat that mistake ever again. That

ancient aphorism "Measure twice, cut once" factors in here, but when it comes to cabinetwork, there is a valuable corollary: Put the perimeter on paper. My first step in preparing plans for the kitchen cabinets was to measure precisely (twice) the as-built dimensions of the kitchen. It didn't make sense to work from the plan drawn up months before. In construction, especially when the designer is also the builder, lots of changes occur that affect the location of elements, usually by a little, sometimes by a lot. It was essential to start with what existed, so my dimensions came not from the plans but from the life-size room itself.

That provided what amounted to a frame into which the puzzle pieces had to be fitted. But how were those pieces to be arranged? If in doubt, look to historical precedent.

In colonial times, the focus was the kitchen fireplace, which had a deep hearth and a beehive oven built into its masonry mass. The cooking was done in pots suspended from an iron crane in the fireplace or on trivets set over coals raked out onto the hearth. Typically, a table was set in the middle of the room, and on the rear wall, a stone sink and a cupboard were mounted. In the Victorian household, the cast-iron cookstove replaced the fireplace hearth and bake oven. Though the kitchen got bigger in the nineteenth century, the arrangement remained much the same.

The advent of electricity meant that refrigerators, dishwashers, and other appliances began to appear. Yet in the years before World War I, the greatest changes resulted from a quest for efficiency. In industry, Henry Ford was perfecting the assembly line; Frederick Winslow Taylor put his stopwatch to work to increase productivity; and Frank Bunker Gilbreth pioneered time-and-motion studies. Gilbreth used motion pictures to study masons at work, examining each operation to identify how a given job could be done in what he called "the one best way."

Best-way thinking was quickly employed in simplifying household duties. In the pages of *Ladies Home Journal,* Christine Frederick described the principles of what she called household engineering. Two of her ideas proved very influential. One was to install shelves, drain boards, and cabinetry as built-ins. Though commonplace today, built-ins were then virtually unknown.

Frederick also sought to identify a systematic approach that would make housework more efficient, just as Gilbreth had done with masons, and Taylor with factory workers. Frederick emphasized what she termed the "efficient grouping of kitchen equipment," but it wasn't until the early 1950s that a group of Cornell University researchers codified what has become the sine qua non of most kitchen designs. The Cornellians called their magic formula the work triangle, though colloquially you will more often find it called the kitchen triangle in design literature.

The notion was the result of laborious time-and-motion studies, but it has a breathtaking simplicity. By observing cooks at work, the researchers recognized that activity in the kitchen converges at three principal points: the sink, the stove, and the refrigerator. In looking at the floor plan of a well-designed kitchen, you can connect the dots between these points and, in most cases, find a triangle. But the Cornell study went beyond description and offered a prescription, an arithmetical formula that sounds like a theorem left over from plane geometry. While it's been variously adapted and amended over the years, in essence the kitchen triangle rule states that the sum of the unobstructed distances from sink to stove to refrigerator and back again should be not less than twelve feet nor more than twenty-two feet. Furthermore, the kitchen triangle rule specifies, no one side of the triangle should be less than four feet or more than nine feet.

The kitchen triangle rule has the aura of received wisdom today. There are exceptions that don't fit the rule, notably in tiny galley kitchens and in giant professional ones where there is a division of labor. And I've certainly heard the little anarchist voice that sometimes speaks quietly in my ear, warning that room should be left for variations that don't quite fit the formula. But the work triangle provided a very handy way to think about our kitchen.

Not surprisingly, then, the sink, the refrigerator, and the cooktop were the first pieces of the puzzle we put in place. Betsy had been shopping, and with our finite budget, she had identified appliances for us. None was a top-of-the-line model, but all were manufactured by reputable makers.

Rather than a range, there would be an electric double oven and a gas cooktop. That way, Betsy explained, you get even heat for baking in the ovens and more immediate control of the burners for sautéing and other rapid operations. As we had resolved much earlier, she ordered two identical dishwashers. We stretched the budget slightly for the refrigerator, as Betsy felt that a model with the freezer compartment at the bottom made sense: since the refrigerator door is opened perhaps five or ten times as often as the freezer door, why not make it easier (and more energy efficient)? Including the washer and dryer units for the laundry, we spent just over five thousand dollars on the seven appliances.

In the finished kitchen, the refrigerator, double oven, and matching dishwashers would assume positions around the perimeter of the U-shaped plan. Between the arms of the U would be an island with a stack of drawers, an overhanging countertop on the far side to accommodate a couple of stools, and, on the near side, the gas-fired cooktop.

Common sense, as well as the kitchen triangle rule, helped establish the locations for these pieces. In roughing in the plumbing, we had centered the risers for the sink beneath the window at the base of the U. That was no innovation but a common design device, since the monotonous routine of dishwashing is made at least a little more interesting by an occasional glance out the window. That seemed almost automatic. The location of the plumbing lines determined the dishwasher positions, immediately to either side of the sink cabinet. That, too, had a certain inevitability about it.

We had framed a large arch to define the mouth of the U. We wanted it to suggest a division between the working area and the rest of the kitchen. Since both the refrigerator and the double oven were deeper than the usual counter width, they could add definition to the arch, anchoring the ends of the U.

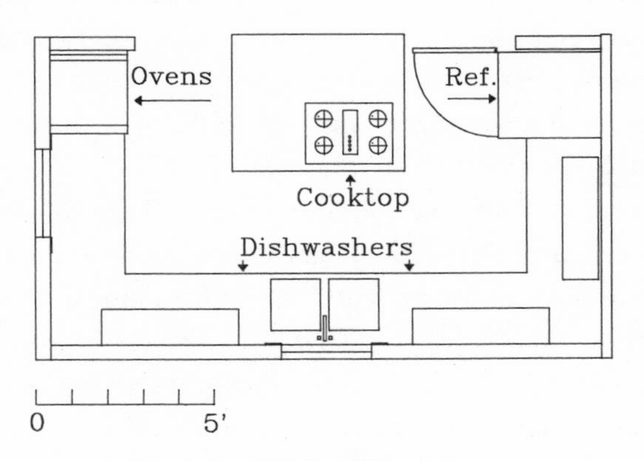

Kitchen Plan

The logic of common practice helped us take the next step. For about half a century, stock cabinets have been built to certain specifications. Countertops are twenty-five inches deep. Undercounter cabinets are two feet deep; uppers, one foot. A three-inch-deep recess at the base of a cabinet (called the kick space) allows the cook to stand at a counter without banging his or her toes on the cabinet base. So-called landing space is recommended, if possible, on both sides of the sink and cooktop. Incorporating each of these good-sense notions, I sketched more lines, and the floor plan suddenly assumed the look of a kitchen.

Still, much of the design was far from predetermined. Consider for a moment that, by common practice, countertops are thirty-six inches above floor level. That makes excellent sense if you're five feet five inches tall (that's Betsy). But take it from me, standard counters seem a little low to a person who's almost six seven. Since our cabinets weren't coming from a factory, we could make them any size we wanted. One evening, Betsy and I found ourselves in the kitchen of our cottage, with our arms at our sides, bent at the elbow, discussing the relative ergonomics of average versus tall kitchen workers. As a food writer, Betsy literally makes her living in the kitchen, so her comfort was the first priority. But she agreed that we could raise the counters a little, perhaps an inch or an inch and a half.

The next morning we found ourselves doing another pantomime, this time in the kitchen at the new house. I had drawn bold pencil lines on the plywood subfloor to indicate where the perimeter counters would be. There was also the footprint of the refrigerator, as well as an arc to indicate the swing of its door. Crude pencil lines represented the double ovens.

The challenge before us was the island. We were balancing several desires: one was for the largest island possible and thus a maximum amount of work space. But the clearance around the island had to be sufficient for the swings of the appliance doors and for one kitchen worker (and sometimes two) to move easily. But the need for clearance had to be balanced with the fact that too much floor space would mean more footsteps *and* a loss of island space.

I had flattened a large cardboard box and cut out a big square, which I had positioned on the floor where the island would be.

Immediately upon seeing it, Betsy said firmly, "It's too big."

"How about if I take six inches off the width?"

"That won't be enough." Her remark seemed a little abrupt.

Ten inches were sliced off, providing more clearance.

The discussion went on: I wanted to center the island on the window for reasons of symmetry. Betsy dismissed the notion. "That's not important."

I looked at her, a little irritated.

She went on. "I want more space by the refrigerator—that, we use constantly. We need just enough clearance by the ovens, since those will be used much less often."

There was more of this push and pull. It dragged on for perhaps a half-hour. Mike appeared, wondering what to do next. I felt a rising sense of unease: Mike and I couldn't proceed with anything else in the kitchen until the cabinets were made, and we couldn't make the cabinets until the plan was finished. The sheets of plywood for the boxes hadn't been ordered yet because that, too, depended upon determining the number of cabinets, how large they would be, and so on. I felt suddenly burdened.

In the middle of our kitchen-planning session—we were by no means finished—I realized that we didn't have a sink. We hadn't decided on its size. In a sense it was the key element that would determine the size and configuration of most of the other base cabinets.

"The sink," I began.

"The *double* sink," Betsy interrupted.

"Yeah, right, the double sink. We need it now."

"You mean I need to go and buy it *today?*"

I nodded.

In a matter of minutes, she was in the car on her way to our favorite plumbing supply house, Mike was putting down the maple flooring in the kitchen, and I was back at my computer. About a half-hour later, though, realization struck: Betsy would find a sink and bring it home, and it would sit untouched for several weeks. The trip was really a waste, as a couple of well-placed phone calls would have gotten the information we needed at that stage.

I had acted out of frustration, sending Betsy off with unnecessary haste. I felt foolish, but I could feel that the days spent designing the kitchen were taking a toll on both of us. I wanted to be back at the work site, to speed ahead, because the deadline was looming; Betsy wanted to be sure the room where she would spend so much time in the coming years was right. In each case, the logic was impeccable.

That evening, Betsy called me away from my computer screen. We'd already talked matters through, and the new drawings I'd produced once we had the sink on hand had pleased both of us, smoothing over our little rift. When she appeared in my doorway after eight o'clock, the time had come to read Sarah and Elizabeth a book. It was a family ritual we all valued, and even during the busiest of the construction times, I happily honored it. I pushed away from my desk gratefully and went upstairs.

Sarah had turned four in early November; Elizabeth was two months short of two years old. Out of some wisdom or instinct, they chose to read *Babar the King*. It was the third of many books about the elephant king, and in it Babar assumes the roles of designer and construction manager of a new city he names Celesteville, in honor of his wife, Queen Celeste. We had read it together before, but that night it seemed richer and more real to me.

The build goes without a hitch (well, maybe that's not so realistic). The elephants move into their new homes in Celesteville. But on the day set aside to celebrate the completion of the new city, two of Babar's valued friends are gravely injured. The Old Lady is bitten by a snake, and Cornelius, the éminence grise Babar relies on, is half-suffocated in a house fire.

Babar has difficulty falling asleep that night. When he finally drifts off, he hears a voice that says, "It is I, Misfortune, with some of my companions." Out his window, Babar sees a legion of grotesque beasts, each of which represents an evil spirit, among them Stupidity, Despair, Cowardice, and Laziness. He is about to shout them away when an angelic host of flying elephants comes into view. They're named Goodness, Intelligence, Courage, Work, and Hope. They and other virtues, the story says, then "chase Misfortune away from Celesteville and bring back Happiness." Babar awakes, dresses quickly, and, upon arriving at the hospital, discovers his dear friends have made a rapid recovery.

By the time we reached the happily-ever-after conclusion, Elizabeth was asleep, and Sarah fading fast. And I felt as relieved as Babar. On that particular night, the story seemed to have been chosen to offer me solace.

The First Christmas

There is always one moment in childhood when the door opens and lets the future in.
—Graham Greene

If the previous months had been an anxious but exhilarating roller-coaster ride, the weeks before Christmas accelerated to a blur. Our deadline was suddenly nonnegotiable, since our cottage was to be sold in the days just before Christmas. Our furniture, clothes, books, and everything else had to be moved, and some twenty members of our extended families would be arriving to celebrate the holiday. The irony was we had so much work that we suddenly seemed beyond worry.

Sarah and Elizabeth were both dispatched to day care Monday through Friday, nine to five, and had baby-sitters for as many hours on weekends as we could arrange. Both Betsy and I felt like bad parents, but we did try to draw the girls into the process. We brought them to the house each day to show them what we had been doing. We gave Sarah a bird feeder for her birthday in November and hung it outside the breakfast-room window at the new house. Within days she spied a chickadee sampling the seed on a Saturday morning visit. Each time Sarah came to the house, she would rush to that window, hoping to see birds feeding there. At first her headlong arrival scared them away, but gradually she learned to approach stealthily. Nuthatches were a constant, and sunflower seeds soon drew purple finches.

As November became December, we stopped showing Sarah and Elizabeth the progress we were making in their bedrooms. One evening, Betsy told them that after dinner we would be going up to the new house so that they could choose their rooms. It wasn't much of a choice, since there were two of them and only two bedrooms, but Betsy's instinct had been right. To them, the prospect was very exciting.

With an air of ceremony, we drove to the house. The four of us walked together to the second floor for a viewing of the rooms. I gave them the facts: one room was a bit larger, but the other was a little closer to Mom and Dad's room. They didn't listen very closely, as both of them were engrossed in inspecting Betsy's paint job. She had painted the rooms soft pastel colors, then returned and applied highlights in two other colors, sponging on splotches to give a mottled texture and a sense of depth.

They checked out the closets, where I had just finished installing shelves and adjustable closet rods at a preschooler's shoulder height. They were amazed that the lights in the closet went on and off when they opened and closed the doors (microswitches had been installed in the doorjambs). Without conflict, Elizabeth chose the slightly larger room, and Sarah, the one that was closer to the master bedroom. When we returned to the cottage, they went straight to their old bedroom but with a new air of expectancy. They had something to look forward to.

THE BATHROOMS WERE ONE step away from being in service after Betsy had grouted the tile and she and Tammy had finished the painting. Setting the toilets and sinks was all that was left to do.

The supply lines were in place, protruding from the walls with stop valves at their ends. The waste pipes were all positioned, too, capped and ready to be connected. So you hang the sink, set the toilet in place, and make the connections, right?

Wrong.

Some assembly was required. The fixtures and fittings came out of their boxes along with bags of miscellaneous parts that had to be identified and fitted together. The faucets and drains, which were purchased and packaged separately, also had to be positioned, caulked, and bolted in place. The tank and the seat had to be mounted on the toilet base; then the entire assembly had to be fastened to the PVC toilet flange that sat flush with the tile floor. The sink in the girls' bathroom would sit on a pedestal but would also be suspended from wall brackets that had to be set at precisely the right height. The antique porcelain sinks mounted in marble surrounds for the master bath and the downstairs half-bath had no bases, so temporary legs had to be devised. They were makeshift affairs, cobbled together with scrap lumber and drywall screws. Later I would come back and make vanity cabinets. Each of these operations was pretty mechanical, but every one required varied tools and supplies.

Once the fixtures were set, the plumbing lines could be tied in. Any threads were wrapped first with Teflon tape; then the nuts were made finger-tight. Supply valves were tightened with a crescent wrench; the waste connections, with a wide-jawed pair of plumber's pumps.

If all the correct parts had been at hand—which they were not—these jobs might have gone like clockwork. Instead, I got a new appreciation for plumber's trucks, the vans or pickups that sit on the job site like self-contained plumbing supply stores. Instead of going only as far as the driveway to find needed fittings, I had to get behind the wheel and drive to the hardware store. That was a very time-consuming part of the process. So was making return visits to trade in the wrong hose for the right one. Or to get another, longer tailpiece, or a 1¼-inch P trap instead of a 1½-inch trap.

If the plumber's truck chockablock with parts is a benefit no amateur can enjoy, the professional has another advantage, too. The pro knows the shortcuts that can be learned only by doing routine installations time and again.

I developed the theory that to be a plumber, a person must have a contortionist gene. Have you ever tried lying on your back, your head and one shoulder jammed into a corner, while you attempt to position a wrench over a nut that you can barely see—only to discover that you have the wrong wrench or that you forgot to put Teflon tape on the threads, meaning you have to untangle yourself, escape from your cramped space, find the forgotten bit, then get back into that awkward position again? Repeat, repeat, repeat.

If the plumber's union were ever looking for a poster boy to convey why people shouldn't be their own plumbers, I might qualify. Still, the bathrooms were fully functional in about two and a half days.

After some further discussion, Betsy and I had resolved our small differences over the kitchen arrangement and I had completed the detailed drawings. With them to guide us, Mike and I put cabinets together quickly, and after the preliminary assembly, Mike joined Betsy and Tammy as a member of the painting crew. I went about the cabinet installation, a job that required care and patience.

The bases were positioned first. Consisting of two-by-fours with blocks between to hold them parallel, each base had to be shimmed to get it precisely level. The boxes were then set on top. Shims were required between them to establish correct vertical positions, with the fronts aligned and the tops level, and to get the spaces between them set precisely so that appliances could later be slid neatly into place. Once the boxes were glued and screwed together, a pine face frame was applied to make the various plywood boxes look as if they were of a piece.

There would be no cabinet doors when we moved in. In our crash program, we did what had to be done, ordering our priorities as we went. That meant postponing many jobs that just couldn't be completed in time. It's the construction equivalent of battlefield triage. In determining what had to be done by our moving date, we established one cardinal rule: all surfaces had to have some finish on them. That meant the walls and ceilings had to be plastered and painted, the floors sanded and sealed, and the woodwork primed. A second rule was that the bathrooms and kitchen had to be in full operation. Cabinet doors; storm windows; hardware such as lock sets, latches, and sash locks; and a variety of decorative touches would come, we told ourselves, in the fullness of time.

Once Betsy had sealed the inside of the cabinets with two coats of polyurethane, I fabricated a temporary countertop out of one-by-twelve tongue-and-groove pine. The front edge had a half-round shape made with the router's nosing bit. The countertop was painted, and the work of installing the sink and the appliances began.

The dishwashers, refrigerator, cooktop, and double oven had been delivered all at once. The double oven was little more than a wiring job. With the aluminum power cable connected, the ovens themselves slid neatly into their cabinet. The refrigerator wheeled in, since it came on casters. Since the cooktop required both a galvanized steel vent pipe and a propane gas line, the crew from the appliance store was better equipped to do that installation. I had to keep saying to myself, *You can't do it all.*

I did opt to install the sink and the dishwashers myself, and that took time. The sink's undercabinet became the point of intersection for two sink drains as well as for the drain hoses from the two dishwashers. The water supply lines arrived and were split to service the dishwashers on either side as well as the faucet above. After some hours lying on my back, reaching over my head and under the dishwashers, I managed to complete the job. No leaks were apparent, but I had a crick in my neck that lasted days and reminded me of what fun it had been.

The cabinet, countertop, and appliance installation took roughly five working days.

MOST PEOPLE HAVE A moving day. We had a moving month.

The operative principle of our mad dash during those last weeks was, *Waste no motion.* That meant a lamp, a chair, or an armful of hanging clothes went along on every trip from our cottage to the new house. As Tammy finished painting each set of shelves in the library-office, boxes of books would arrive. With their contents placed on the shelves, the boxes would get tossed into the minivan to be returned to the cottage for the next load. We were so consumed with the work of finishing the house that we didn't have the luxury of taking a few days to organize a logical move of our worldly goods. Every night we would pack a few boxes, and the next day they would be moved.

We decided we needed a grace period between the time we actually went to live in our new house and the day we had to be out of our old one. That way, if something was amiss, we could always return to the cottage for a night or two. Our first night in the house was December 15. Having left the girls with Miss Diane at Long Barn Daycare for the night, Betsy and I worked until nearly midnight making our room habitable. We fell into bed, exhausted, but even on our first night in the house, we felt as if we belonged there.

In the morning, we got up and used our new bathroom. Everything seemed to work. When we went downstairs, the kitchen was a mess, but we managed to have juice and cereal. It felt like home, except for the absence of Sarah and Elizabeth.

We lit the grubka. I had a few misgivings (which I'd kept to myself) about how well the smoke would be drawn through the maze of up-down-and-across flue lines. In moments my doubts were forgotten. A single ball of newspaper initiated a strong draft. Betsy watched as I added a few bits of kindling (we had a cellar full of wood scraps), and in less than a minute we were listening to a crackling fire. We had only a small fire that morning to be sure the masonry was fully cured, but the ease with which that first fire had lit was very encouraging. In time, we would gradually test the output of the masonry heater.

That day we got the girls' rooms organized. Mike and I relocated power tools to the cellar. We moved furniture, too, arranging the living room. In the evening I hung a mirror and a couple of pictures. We prepared a simple meal of pasta, and both girls seemed as happy to be there as we were.

In the days that followed, Tammy kept painting, and Mike and I made the stops for the windows, pieces of vertical trim that sat inside the jambs and held the windows in place. There were innumerable little jobs to be done, and all hands were kept busy.

Two days before we were to sell the cottage, the building inspector made what we hoped would be his last visit. Previously he had inspected the foundation, the insulation, and the rough-in of the plumbing. On his several visits, he had found nothing that required changes, but with the clock ticking down, I was nervous.

No sooner had he gotten out of his car than he pointed to the four steps leading to the front door. "More than two steps means you gotta have handrails," he said. My fear was that this would be the first of a long list of citations.

We were lucky. "Put on a rail," he said on his way out a few minutes later. He had to step around Betsy, who was kneeling on the floor of the mudroom, setting tiles. "Then I can give you your C of O."

In about two hours work, we knocked together some sturdy (if temporary) railings, and we had our certificate of occupancy.

As the sale of the cottage approached, Betsy packed up the attic, and Mike and I brought load after load of boxes to the new house. A couple of trips to the dump were required. We found ourselves asking repeatedly, "Why did we save that all these years?" Our lawyer and the real estate agent handled most of the details, so the closing came and went painlessly, involving little more than a few dozen signatures. The most momentous event of that day proved to be Betsy's discovery of our Christmas tree stand. That prompted a break in the action.

There are people who think the past isn't important. "I live in the present," they say. "I don't want to be weighed down by all sorts of old stuff." Often that means objects—like old photos, family furniture, or the school papers your mother set aside for you from elementary school. Some people want to forget, but I want to remember both childhood and parenthood.

I grew up comfortably, with a creative mother and a couple of much older brothers who were very much available to me. My father was mostly absentee, drawn by his work to Boston, more than fifty miles away, working most evenings and weekends. While he wasn't what we today would call child-friendly, he was on balance a decent father, but I knew his laissez-faire mode of parenting would not be mine. From the moment of Sarah's birth, I made a point of being there and in real contact with my children. Ironically, though, as Christmas approached and we moved into the house, I had become almost as inaccessible to my children in the preceding weeks as my father had been to me.

One opportunity to change that offered itself when Betsy appeared with the tree stand in hand. She suggested that, as a family, we go and chop down our own tree. In previous years, Betsy had bought trees off the rack, precut and shipped to the point of sale. But the first Christmas in our new home seemed a suitable time to establish a family tradition. We decided to harvest a tree that very afternoon.

After lunch, Betsy had gone off on errands, and she was late arriving home. Having quit early to be with the girls, I got Elizabeth up from her nap and put her in Betsy's and my bed along with Sarah. We read first one book, then another. While I enjoyed this rare afternoon treat, my anxiety level rose as four o'clock approached. A snowstorm was predicted for that night, and we needed to get on with the tree-cutting in order to be able to pick a tree that wasn't covered with a thick layer of snow.

When Betsy did arrive, at a few minutes after the hour, we hurriedly bundled the girls into their snowsuits and car seats and set off. The destination was a tree farm three miles away. It had been an open field perhaps twenty years earlier, but each year its owner had planted a few rows of evergreen trees. Betsy had seen him there cutting trees and had gotten permission for us to pick our own. "Just put twenty dollars in the toolbox on the back of the tractor," she had been instructed. As we pulled carefully into the icy entrance to the tree lot, the tractor was visible, partly hidden by a large blue spruce.

The sun was dropping. I grabbed a handsaw and a pair of pruners. Sarah carried the clothesline we would use first to drag the tree, then to tie it onto the roof of our minivan.

We walked into the tree farm. Well, three of us walked, as Elizabeth soon insisted on riding on my shoulders. We headed in the direction of the disappearing sun. In the gloaming, it was a black and white world, as everything was a shade of wintery gray.

We stepped clear of a row of overgrown Christmas trees, facing directly west. The sky had become crimson as the sun dipped behind the Red Rock ridge. The thin layer of snow took on a pinkish hue, reflecting the scant surviving light. I helped Elizabeth off my shoulders, and the four of us fanned out.

We hurriedly examined the silhouettes of trees. It was a race. The light seemed to be fading by the second. Elizabeth tripped over a branch and needed consoling. We wanted the right tree, and we wanted it right then. One tree looked straight and shapely, but Betsy wasn't sure. We looked at others but soon returned to the first selection. The girls were both enthusiastic, and Betsy finally agreed. "Let's take it," she said confidently.

With a handsaw I cut the tree down and called out, jokingly, "Timber!" It was neither tall nor heavy, and though Sarah was well out of range, her face at that moment was a mix of fear and joy.

We stumbled back to the car in the dark. Betsy and the girls warmed themselves inside while I tied the tree to the roof rack by the light of a flashlight. My hands were stiff with cold, but there was an air of expectation as we drove home. When the tree was upright in our living room a half-hour later, we discovered that the top was less than a quarter inch from the ceiling. It was obviously the right tree.

Our little adventure had been hurried, but traditions are sometimes born hard. Firsts are important: the first word, the first step, the first day of school, and the first Christmas tree in a new family home.

CHRISTMAS MORNING ARRIVED QUICKLY. Betsy woke up before first light, worried about getting the turkey in the oven. For my part, I had a punch list of things to do before company came. One of those things was to install a latch on the downstairs bath so that our dinner guests would be able to keep the door closed.

Betsy added another task to my list: starting a fire in the Rumford fireplace. That was a pleasure. It took only a moment to open the damper, and the draft was instantaneous and strong. The welcoming crackle of the fire was soon to be heard. The fire screen we had brought from our other house didn't really fit the Rumford opening, but I told myself we'd find a suitable one soon enough.

By early afternoon, the house was full of people. Members of my family arrived, having traveled from California, Vermont, Massachusetts, and New Jersey. Several from Betsy's family appeared, having come from the state of Washington. New York City friends arrived by train.

We opened presents in front of the Rumford's blazing fire. I explained the workings of the grubka perhaps seven times to various guests. The dining room was tightly packed, but no one seemed uncomfortable. Betsy had made the dining room a target. "I know the house won't be truly finished for Christmas," she had told me, "but we need at least one room that's civilized." In the week before Christmas, she had laboriously painted and repainted the cornice, chair rail, and baseboards a deep Venetian red, a color that was popular in the early nineteenth century. We had found a source of reproduction paints that employed the same raw materials that were used before the Civil War. The pigments (iron oxide in the case of Venetian red) were ground by hand and mixed with a base of linseed oil.

Betsy found an inexpensive chandelier to suspend over the table. Its finish was faux verdigris, with four arms that held small frosted-glass spheres. She devised a way to give them a rosy hue, brushing thinned splotches of the Venetian red on the frosted glass. On Christmas Eve, she had hurriedly made swags for the windows. In the waning afternoon light on Christmas Day, with a dozen candles burning, the room had a timeless feeling.

Most people left that evening, but some stayed overnight, a few for several days. The company helped cushion the shock we felt. We had been going full speed for so long, and like runners in a race, we didn't want to stop abruptly. Having family and friends in the house allowed us to slow to a trot and then to a walk.

Surprisingly, as we coasted through the holidays, there wasn't a sense of anticlimax. We were tired, and anyway, we had a new job to do: to establish our patterns, our way of life, in our new place.

EPILOGUE

The World As We Would Have It

A stranger on the indifferent earth,
[man] adapts himself slowly and painfully
to inhuman nature, and at moments, not without
peril, compels inhuman nature to his need. . . . He may
cower before it like the savage, study it impartially for
what it is like the man of science . . . [or] he may
construct, within the world as it is, a pattern
of the world as he would have it.
—Geoffrey Scott

The house has been finished for several years. Betsy would add a qualifier to that statement—it's *mostly* finished—but in truth the place is as finished as most houses ever are, as we continue to add details, fix problems, and perform routine maintenance.

There are countless signs of jobs done in a rush. There are molding joints that are a bit sloppy. There are many examples of hurry-up workmanship where one more cut, another coat of paint, or a second sanding would have made for a better surface or joint. There just wasn't time.

Since we moved in, more changes have been made outside than in. Mike and I built a three-car garage with a workshop above it a couple of years ago. Betsy has established a serpentine perennial bed that, over the years, has stretched to more than one hundred feet in length. It follows the undulating line of the ha-ha, poised on the brink of the tall stone wall. The rank and file of flowers stand guard, a wall of color and pattern that marks the division between the manicured lawn and the rougher, wilder meadow beyond.

We planted trees, and some have thrived, while others have died. A couple of the ones that survived were planted too close to the house. That was an error of scale and anticipation. An asparagus bed, now six years old, provides generous quantities of asparagus in May and June.

Elizabeth shows signs of being a natural gardener. She's not a weeder yet, of course, but she loves to pick flowers and harvest vegetables. I have a happy image in my mind of a day when Mike, who still works for us on an occasional weekend, was moving rich topsoil for a new garden. Elizabeth was happily walking alongside his wheelbarrow, "helping" him do his work. Both were quite evidently enjoying the experience.

Sarah and Elizabeth are my partners in a small kitchen garden where we grow lettuce, peas, beans, tomatoes, and other vegetables. Sarah loves to swim in the pond and emerges from its muddy waters looking like a swamp creature.

We stocked the pond with fish, eight- to ten-inch rainbow trout from a local hatchery. The plan was not to fish for them but to use them to help establish the pond as a healthy ecosystem. But the day after we emptied a cooler full of forty trout into the water, the fish did become prey.

The big bluish gray bird hadn't been seen for two years, but as if by invitation, the heron returned. The bird that had seemed so magical became instantaneously a pest, and in a matter of days, it ate most of the rainbow trout.

Mark graduated and went on to become the news editor for a Web site in Oxford. He came back to Red Rock one January and experienced the winter wonderland he came for. He had never seen more than a fraction of an inch of snow before, having known only the wet, quick-to-melt stuff in the United Kingdom. When he arrived here, the snow was falling steadily, and by the time it stopped in the night, a total of eighteen inches had accumulated. It all fell gently and quietly, with a minimum of wind.

In the house, we remained warm, thanks largely to the grubka. Unlike a modern appliance, the grubka came with no owner's manual. It heats up slowly, requiring twelve hours or more for its exterior brick surfaces to reach optimum operating temperature. Then the brick stays consistently warm—about 110 degrees—around the clock.

Ralph came by for a visit one winter day, bringing with him his son, Dominick. In his early twenties, Dominick had joined the business, raising the total to four generations of masons in the Bruno family. Ralph had yet to build another grubka, but he studied ours carefully, asking about its function. I told him we were happy with the device, and indeed we are. It has become a ritual part of cold-weather life. When we light a fire in it for the first time each autumn, it is almost a celebration, more real than a calendar date like the first day of winter, a tangible occurrence that connects us to our environment.

In the years since we moved into the house, both business and pleasure trips have taken me away from Red Rock. Before leaving, I have gotten into the habit of hiking in the woods, walking the same circuit we first walked on the day Betsy spied the mayflower. The exercise is good preparation for the sedentary hours to come riding in an airplane or a car. The walks represent more than that, too, enabling me to take along a fresh sense of our home, whatever my destination.

The excuse of going away isn't required for a stroll through the woods. I have taken the same walk a thousand times in the last few years, often alone, but regularly with Sarah or Elizabeth. My daughters no longer need to be carried but often run ahead along the trail or on self-guided detours into the woods. Elizabeth is a natural magpie and invariably returns to the house with bits of bark, feathers, or wildflowers that she has found along her rambles.

There must be something instinctual about stone moving. In low-lying areas on my path where standing water accumulates after a heavy rain, I have dropped boulders into the silty earth to establish stepping-stone paths. There's a cairn down by our pond, too, and there's no mystery about its origins. On my walks through the woods I have carried suitable flat stones back to the edge of the pond. Over a period of several years, the stack of shale has gradually grown.

I usually make just one trip around the loop; it takes about fifteen minutes. Yet each visit offers something different. There are animals to see, ranging from twittering squirrels and woodpeckers to deer and turkeys. Orange newts dot the path after a rain. One day I brought a few home in a small matchbox. Although initially afraid, Sarah and Elizabeth played with them for an hour or more on the front step of the house.

At times I see evidence of darkness in my woods. One day on my walk, I stumbled over a short chain. One end was nailed to the base of a large hickory tree; the other, attached to a small leg trap. The jaws of the trap were not set and hadn't been for many years, judging from the thick crust of rust. No doubt it belonged to Charlie or even his father, both of whom had trapped raccoon and foxes. Animal pelts were once an important source of cash in Red Rock's tight rural economy.

Another day, Elizabeth and I found that two tall hemlock trees had come down across our familiar path. Finding the downed trunks was no surprise; what was remarkable was the way the trunks of the trees had cleaved fifteen feet above the ground. The only logical explanation was that the dense network of needles at the treetops had caught a strong wind, like a sail, while the lower branches, shielded by a ridge, had remained unbowed. While one top had fallen to the ground, the other, in a million-to-one fluke, fell onto the jagged trunk of the first. The pointed tip of the standing trunk had impaled the falling treetop like a bayonet.

It was an extraordinary sight, but Elizabeth, four at the time, immediately chimed in, "Look, Daddy, it's a T tree." What to me was a striking example of nature's unpredictability and power was, to a precocious child, just another of life's little alphabet lessons.

I KEPT NO DAY-TO-DAY records of the time it took to build our house, but my best estimate is that I spent about four thousand hours in the sixteen months between August 1993, when we broke ground, and December 1994, when we moved in. In the months before construction began, a great deal of time was invested in designing, planning, estimating, and getting ready to build. In the years since, we have gone back and completed unfinished jobs. My hours probably approach a total of about five thousand; Betsy's, one thousand. That amounts to about three years of forty-hour weeks. The final cost was a bit under seventy-five dollars per square foot, roughly the national average for all houses. Building contractors in our area would have charged well over a hundred dollars per square foot, probably much more, for equivalent quality. We traded time, sweat equity, for a finer house than we could otherwise have afforded.

Such numerical calculations convey little about the nature of the work. One of the joys of building is that it is at once work and leisure. Physical activity is a human pleasure: using the muscles in such pursuits as sports, walking, and gardening (or building a house) enhances appetite, promotes sound sleep, and contributes to an overall feeling of well-being. Physical conditioning enables us to look in the mirror at healthy skin and good muscle tone. Then, too, there are the sensations of perspiration and happy tiredness.

That's the body stuff, but there's a brain component, too. Building is not mindless. Only occasionally is construction confined to repetitive tasks like ditchdigging or running a machine to perform the same operation many times in succession. Building is a varying mix of jobs, some of which require strength, others finesse. There are situations that tease the brain, demanding precise calculation or planning. Other jobs are all instinct and leaps of faith.

The human animal is unique in its ability to delay gratification for months or years at a time. That is what Betsy and I had to do. What we did in presuming to build a house was act out a dream that many, many people have. If building a house is the ultimate do-it-yourself project, then the highest calling of a do-it-yourselfer is the ability to see something beyond the vanishing point.

In that sense, building a house was a great lesson in managing expectations. The ultimate gratification is to move in, but that may be a long way off. So you learn to take pleasure from the passing satisfactions. Managing expectations is a valuable life skill. Even if your goal is out of sight, it cannot be allowed to be out of mind: it must be there, dimly imagined or richly attractive. We all in some way practice the delay of gratification, like the child who eats the chocolate wafers of an Oreo first in order to enjoy the icing inside in a single bite. Our house was simply a much larger sweetness we wanted to experience.

We discovered another pleasure along the way: a deep satisfaction comes from focusing on a long-contemplated goal and, in the process, enjoying the means of reaching it. Perhaps the experience is like that of the athlete who, through years of practice and imposed discipline, comes to focus on the competition, or of the writer or artist laboring in obscurity, "paying dues," waiting for the break. The struggle becomes worthwhile for its own sake.

Building our house was like that. A tiny, almost inexpressible idea was conceived years before we had a clear notion of what we would build, not to mention where we would build it. The gestation required more time, as we determined what we wanted and needed and what it would cost. Then there was the doing of it, the daily getting out of bed, walking to the work site, buckling on the tool belt, and then working. On any given day, a nagging backache, the common cold, or just plain tiredness in the morning could not be allowed to interrupt the momentum. The background noise of life—details like bills, stresses such as an overdue magazine piece, or lingering regrets over the mistakes of the past—gradually faded away. No drill sergeant stood by to order compliance. The will had to come from within, and it did.

In short, I do not regret the time invested in the build. If I had not done this thing for myself and my family once we conceived it ten or more years ago, there would be a sense of regret at a path not taken. Yet expectations can be surprising. You think that the big game is everything—you must win or it's all wasted. You must have that job, win that prize, get somewhere first. Then you don't (and someone else does). Strangely, over time, you realize that the failure wasn't tragic. Other challenges, opportunities, other goals, come along. And, oddly, you may discover that the other guy, the "winner," didn't find the "prize" to be the one-way ticket to nirvana he—and you—thought it would be.

I'm proud of what we did. But the best part was doing it. Not that moving in was a disappointment. It wasn't, and I very much like living here in the house where I am writing this. But much of the pleasure was in the execution.

There's satisfaction, too, in hearing other people say admiring things about the house, but the stage when we felt the need to point things out to people has passed. If asked about the house and its genesis, we will tell people as much as they want to know, but if they're not curious, that's fine, too. A friend recently brought a guest of his to visit. The stranger evinced an interest in some of the details of the house, so we went on a little tour. At one point he asked if we had much trouble finding good contractors. A bit disingenuously, I told him no, the guys we'd hired had been terrific. He left unaware that we had done the bulk of the building ourselves. On the other hand, I still get a little charge when the penny drops and a visitor asks, "You mean *you* designed and built this house?"

We got a visit our first winter from an insurance appraiser. He didn't knock on the door, but examined the exterior of the primed but not yet painted house, measured its perimeter, and took a couple of photographs for his report. We knew nothing of his visit until our insurance agent called. An old and valued friend, Glenn was laughing.

"Just got your appraisal," he reported. "Let me read it to you. 'Nineteenth-century home, in midst of renovation.' "

I WAS PREPARING FOR a business trip when I heard Charlie Briggs had had a heart attack on the operating table during bypass surgery.

I telephoned his wife, Helen, who told me gravely that he was alive but on a respirator. When asked if he could have visitors, she said, "It's supposed to be family. But we'll tell them you're his son." She laughed at that—he had no son—and I imagined her small chuckle was one of the few light moments of her week.

"I'll go tomorrow," I told her. *On my way to the airport,* I told myself.

The next morning I awakened sneezing and sniffling, a cold having come on in the night. Charlie already had fluid in his lungs, Helen had told me, and no convalescing heart patient needs to be exposed to the common cold. My visit would have to be postponed until my return.

The next and last time I saw Charlie, however, was in his casket. He died a few days after his surgery.

Charlie had his eccentricities, but they were outweighed by a central strength, a positivism, a set of convictions. When a house in Red Rock had burned in 1941 for the lack of a nearby fire engine, he founded the fire company and served as its first president. When the dwindling congregation at the Red Rock Methodist Church could no longer support a minister and joined forces with a nearby church, Charlie founded the Red Rock Historical Society to save the church building.

At the end of his life, his complaints had begun to get the best of him. I remember going to his house to deliver a quarterly mortgage check. The house was dim, and thinking no one was home, I decided to leave the check where he would find it, on his desk, just inside the door.

Reaching for the door knob, I saw a movement out of the corner of my eye. A shadowy figure rose from a chair. It was Charlie, who had been sitting in a nearly dark room, alone with his thoughts.

Almost surely, depression sat heavily on his shoulders. He had little life of the mind so as his body failed him, he did not turn to books for company as some people would have done. In that sense, Charlie was hardly a historical person. His desire to convert the Red Rock Methodist Church to a historical society had little to do with history by any academic definition, but his sense of community was historical. He was someone who had rarely left Red Rock and had seen little of the larger world, and his memories were of his community. Charlie was a man with no presence beyond the narrow confines of this hamlet, yet he was a man whose essential worth transcended this obscure place.

When Betsy and I came home from Charlie's funeral, we picked up Sarah and Elizabeth from day care. Elizabeth was her blithe, lighthearted self, but Sarah's finely tuned antenna picked up the emotional tension. She asked questions like, "Where did you go today?" She was fishing but wouldn't be fooled by less than the truth. The suit and tie I wore, unaccustomed attire for me, begged a question, too.

Shortly after we got home, I took her for a walk in the woods. Not entirely by plan, we walked into the potato patch. As we emerged from the dense woods into the open field, I told Sarah that Charlie's dad had grown potatoes there many, many years before. Since Charlie had always called it the potato patch, so did we.

She asked me if Charlie's father was dead. We had told her Charlie was, so that knowledge, along with our visit to the potato patch, cued the question. I told her that he was, that he had died a long time ago, before I was born.

I told Sarah about Charlie's funeral, about how several people she knew had stood at the front of the church and talked about Charlie. She asked what they said, and I told her quite honestly that, mostly, they had said he was a man who had always tried to do the right thing.

She took it all in silently, and we turned onto the path for home.

As we walked, Sarah asked me, "Was Mr. Briggs there today, too?"

"Yes, but he was in the box they were burying him in."

She accepted the answer with a nod.

One of the challenges of parenting is telling the truth in just the right measure. Betsy had been concerned that talk of an open casket wouldn't be age-appropriate and could upset Sarah, but Sarah asked, "Was his container like Snow White's?"

I almost laughed but held it in. Thanks to Walt Disney, open caskets obviously weren't unknown to her.

"Yes, it was a little like Snow White's—"

"You mean you could see him sleeping in there?" she interrupted.

I explained that, no, he wasn't sleeping, but that he had died and that was different. But yes, the box was open.

Sarah, like all curious children, was groping through the wide expanses of life she had no experience with in order to fit new pieces of it into her focused worldview.

She asked, "Is everybody buried in an open box?"

I explained that they closed the casket before they buried it.

"Will you be buried in an open casket?" she asked.

"No, I don't think my casket will be open. I think I'd like to do as my dad did, your grandpa. When he died, he didn't want to be buried in a box, so he was cremated. That means his remains were put in a giant oven, box and all, and reduced to ashes. Then we buried his ashes in his lily garden, which was his favorite place in the world."

After absorbing that, she asked, "Where will you be buried?"

I told her I wasn't planning on being buried for a long, long time. She interrupted, a bit impatiently, saying, "I know, I know."

Then I heard myself saying, "I want to be buried here in Red Rock. I don't know where in Red Rock, Sarah, but this is our home."

That was enough for her. She was getting tired, so I put her on my shoulders and soon we emerged from the woodland path.

As I always do, I experienced a little thrill upon approaching the clearing where our house is set. First I glimpsed the pond; then the ha-ha became visible through the screen of trees at the edge of the woods. The house seemed to materialize, the large expanse of its front reflecting the bright sun. I spotted the frozen, craggy canes of the wild blackberry bushes that had simply appeared at the base of the great stone wall. The bushes made me think of Charlie; the wall, of Robbie; the house, of Palladio. I thought, too, of Mark and Mike.

Coming home is a pleasure common to most people. But the sensation is all the more acute for people, plain or fancy, who hatch an architectural idea; nurture its sticks, stones, and surrounding terrain; and then go to live in it with their families. Here in the Red Rock wood, Betsy, Sarah, Elizabeth, and I are in our place, in the world as we would have it. For however long we get to live here, we are indeed fortunate.

Floor Plans

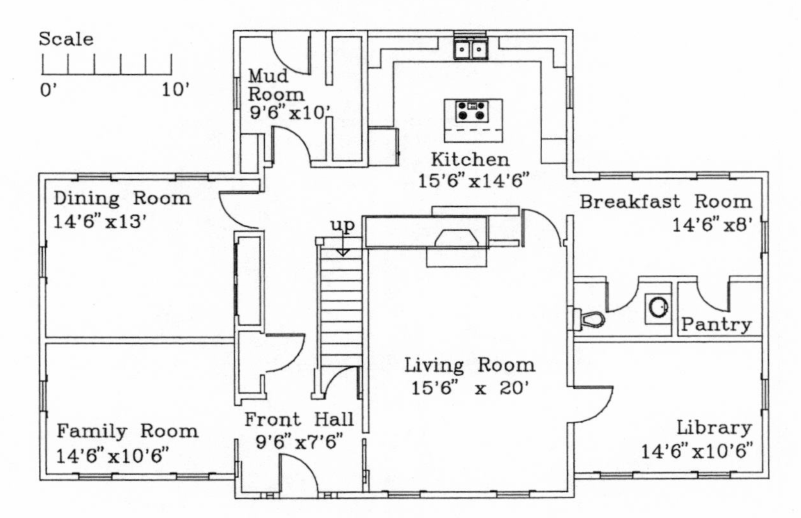

First Floor

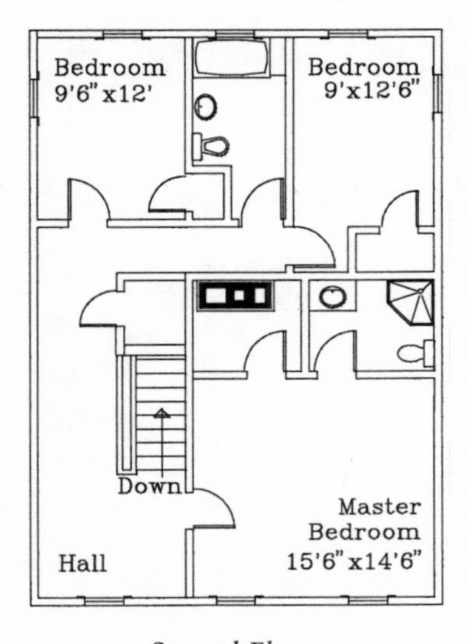

Second Floor

Framing Plan

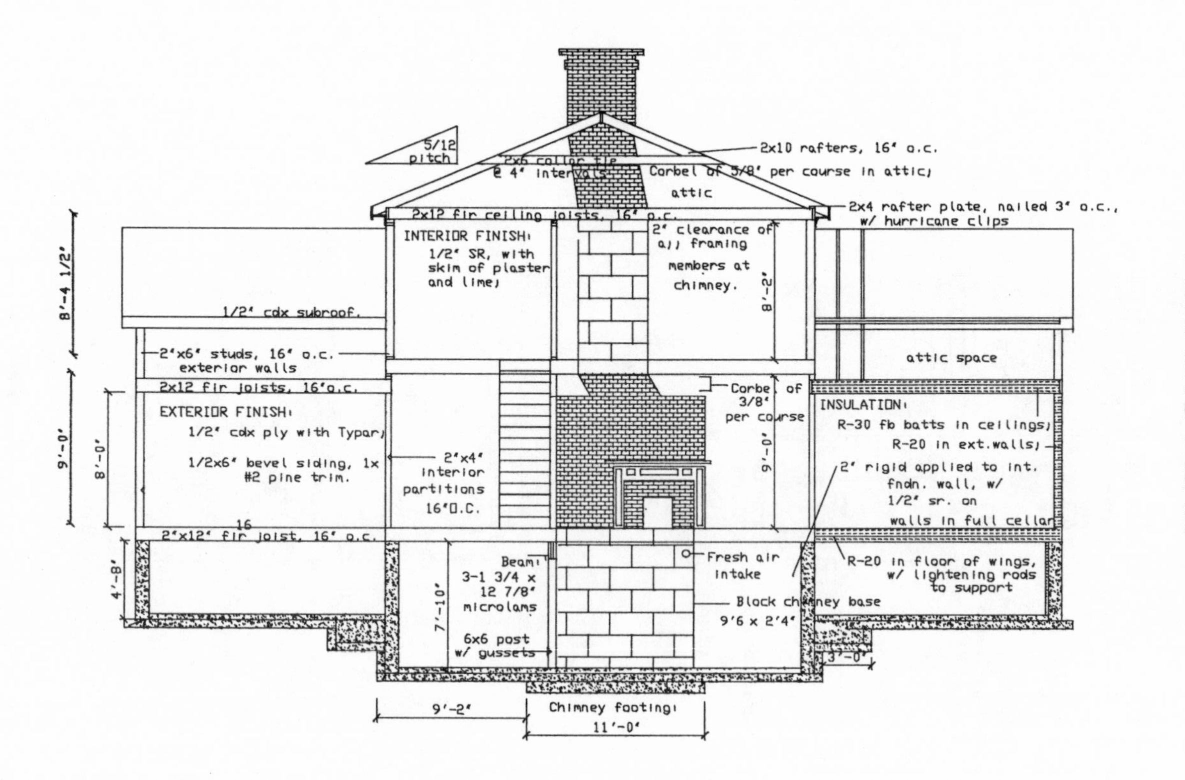

NOTES AND ACKNOWLEDGMENTS

The story told in the preceding pages is true. Since real life doesn't always offer the convenience of fiction, however, I had to foreshorten the process in writing it down, selecting representative events in order to produce a manageable narrative. As a result, I cannot pretend that this book recounts every moment of the build; on the other hand, in writing it, I drew upon notebooks, calendars, receipts, and other records, as well as recollections, and made every attempt to remain faithful to the facts, dates, costs, and all other particulars of the building of our house.

The characters that appear in this book are real individuals. None is a composite, and with only two minor exceptions, no names have been changed. Once again owing to considerations of length, not every contributor to the process could become a character. Thus, there are friends, relatives, antiques dealers, contractors, suppliers, and a range of others who helped us build our home but go unmentioned in the text. In the notes that follow, I have tried to cite those who, sometimes unknowingly, contributed. We also owe a debt to a variety of architectural, literary, and historical sources, and references to the most important of those are cited below as well.

This book, as distinct from the construction process it recounts, benefited from the guidance and goodwill of more than a few people. Special thanks and acknowledgments must go to Dominick Abel, my agent, who has been with this from the start, offering wise editorial commentary and great patience with the pace at which the manuscript developed; Antonia Fusco, my editor at Algonquin, who has done what only the best editors do—made the book better, made its author think and work harder, and directed the process with wit and kindness; Elisabeth Scharlatt, the publisher at Algonquin, who has brought great enthusiasm and energy to the publishing process; my valued friend and collaborator Kathleen Moloney, who read and critiqued the manuscript at a time when a fresh eye was badly needed; Rachel Careau, for her exacting and subtle copyediting; a favorite traveling companion and sometime collaborator, the photographer Roger Strauss III, who offered encouragement as the book took shape and told me the story of the Vitruvian day spent on his house site with Charles Gwathmey in Purchase, New York; and others at Algonquin who have added immeasurably in the challenging job of reaching the audience for this book, including Andra Olenik, Dana Stamey, Shelly Goodin, and Maggie Laws.

As to the house itself, Betsy's and my first thanks must go to Mike and Mark. The premise of this book is that I designed and built a house for my family. I may have put in more man-hours than anyone else, but Mike Beecher ran a close second, and Mark Lynas came next. To them, we owe an immeasurable debt. They gave of themselves, and because of their strong backs, good hearts, and generous spirits, they will forever be welcome in our home. As will Robbie Haldane, whose ability to read the future of a landscape is nothing short of remarkable. And, of course, our appreciation must go to the late Charles Briggs Jr. and his widow, Helen, neighbors and friends.

In more or less chronological order, I also wish to thank and acknowledge the contributions of the following:

In chapter 1, "The Footprint," George Lee, who, acting as an intermediary, helped us reach an understanding for the purchase of land from Charlie; Pope Lawrence, who staked out the house with me on a hot summer day; Vitruvius and *The Ten Books on Architecture* (New York: Dover Publications, 1960) for much wisdom, in particular about siting the house; Lee and Dave Madsen, of Madsen and Madsen Concrete Contractors, who, together with their crew, put in our foundation; the *Chatham Courier* and its April 10, 1975, piece reporting Charlie's nonappearance on television; building inspector Marty Ritz, for his reasonableness; and W. Roger Goold, of Gordon W. Goold, Inc., for his recollections of the 1934 Chevy.

In chapter 2, "Building the Box," Jean Atcheson for delivering us her great-nephew, Mark; Ed Herrington, Inc., purveyor of building supplies from which all the red trucks arrived; and Connie Weissmuller at Tor House, our guide on that first visit and to things Jeffers ever since, together with John Hicks, the late Lee Jeffers, and James Karman and his book *Robinson Jeffers: Poet of California* (San Francisco: Chronicle Books, 1987).

In chapter 3, "The Secondary Imagination," John Fyler, professor of English, for helping me appreciate the mystic Coleridge; Edgar Tafel and his memoir, *Years with Frank Lloyd Wright* (New York: Dover Publications, 1979), for the Fallingwater story; Glenn Williams, for introducing me to Philip Palmedo and the story of his house; Professor Duncan Stroik, for his insight into the primitive hut in general and for directing me in particular to *An Essay on Architecture* by Marc-Antoine Laugier (Los Angeles: Hennessey and Ingalls, 1977); John I Mesick, from whom I seem always to be learning something new about old buildings; John Harris, for his book, *Sir William Chambers* (University Park: Pennsylvania State University Press, 1970); and *Great Georgian Houses of America,* Vol. II (New York: Scribner, 1937), wherein I first saw the Nathan Roberts House in Canastota, New York.

In chapter 4, "The Matrix Materializes," Joe Iuviene, R.A., whose critique of the plans for our house made it a better structure while respecting our design notions; Suzanne Walker, friend to each of us; and Jeremiah Rusconi, a most sophisticated student of old buildings, who sold me the French doors and, later, the marble sink surrounds.

In chapter 5, "Of Hearth and Home," Donald Carpentier Jr., who offered much general guidance but also directed us to our staircase and doorway, as well as to Ralph Bruno; *The Forgotten Art of Building a Good Fireplace* by Vrest Orton (Dublin, New Hampshire: Yankee Books, 1974), a diverting though highly polemical reexamination of Rumford and his fireplaces; and *The Book of Masonry Stoves* by David Lyle (Andover, Massachusetts: Brick House Publishing, 1984), the bible for the construction and history of masonry heaters.

In chapter 6, "The Negative Moment," Chris Carozzo, who, along with Mike Beecher, helped to roof the house; Diane and Braxton Nagle of Long Barn Daycare, for creating a perfect home-away-from-home for our children; and Linda Waggoner, the director at Fallingwater, for telling me of its "negative moment."

In chapter 7, "A Winter's Work," Jerry Grant, for his friendship, frequent counsel, and help in dismantling the stairs; Fred

Engel Heating, Cooling, and Plumbing, our HVAC contractor; Wayne Walker, a fine companion since first grade, who helped with many tasks on weekend visits, including the move of that giant entryway, along with Jim Krasawski; antiques purveyors Zane Studenroth Jr., Jerry Jordan, Russell Carlsen, Robert Herron, and Steve and Sue Anderson, for sundry house parts; and, again, Steve, for so often sharing his knowledge of building, both firsthand and anecdotal.

In chapter 9, "An Imagined Arcadia," Jim and Nancy Foran, the absentee neighbors who allowed me to be so free with their hillside; Deborah McDowell, to whom Robert Haldane told the mouse story, and her article "Tell It to the Stones and They Will Listen" (*Irish America Magazine,* November 1991); James Ryan, the late but warmly remembered longtime site manager at Olana, as well as Karen Zukowski, former curator, and Heidi Hill, director of education; Albert, Matt, and Greg Verenazi, who, together with Leo and Felix Gardina, William Warner, and Tim Stalker, piloted dump trucks, backhoe, excavator, payloader, tractor, and dozers in the process of creating the green expanse that surrounds the house today; and Simon Schama, for his learned and lively *Landscape and Memory* (New York: Alfred A. Knopf, 1995).

In chapter 10, "Of Columns and Completion," Robert Tavernor, for his enlightening conversations, for his excellent introductory volume, *Palladio and Palladianism,* and for his and Richard Schofield's new translation of *The Four Books on Architecture* (Cambridge: MIT Press, 1997); Bruce Boucher, for his readable and luxurious *Andrea Palladio* (New York: Abbeville, 1994); and other writers on Palladio too numerous to mention here.

In chapter 11, "The Race to the Finish," Mihaly Csikszentmihalyi and his book *Flow: The Psychology of Optimal*

Experience (New York: Harper and Row, 1991); Christine Frederick, whose *Ladies Home Journal* pieces were later collected in her book, *The New Housekeeping: Efficiency Studies in Home Management* (Garden City, N.Y.: Doubleday, Page, 1913); Kevin R. Lynch, of Hardwood Flooring of New Lebanon, New York, who did the sanding; and Jean de Brunhoff, the author of *Babar the King* (New York: Random House, 1935).

And to the many other friends and family who during the build offered support, commentary, occasional labor, and goodwill when each was needed.